78 Structure and Bonding

Bioinorganic Chemistry

With contributions by
E. Bill   E. L. Bominaar   O. M. N. Dhubhghaill
C. Friesen   B. K. Keppler   H. G. Moritz
P. J. Sadler   A. X. Trautwein   E. Vogel
H. Vongerichten   H. Winkler

With 86 Figures and 34 Tables

Springer-Verlag
Berlin Heidelberg GmbH

ISBN 978-3-662-15013-9

Library of Congress Cataloging-in-Publication Data
Bioinorganic chemistry/with contributions by E. Bill . . . [et al.].
p.   cm.—(Structure and bonding; 78)
ISBN 978-3-662-15013-9   ISBN 978-3-540-47535-4 (eBook)
DOI 10.1007/978-3-540-47535-4

1. Bioorganic chemistry.   2. Organoarsenic compounds.   3. Iron proteins.
I. Bill, E. (Eckhard), 1953–   . II. Series.
QP531.B54   1991   574.19′214—dc20   91–24733CIP

© Springer-Verlag Berlin Heidelberg 1991
Originally published by Springer-Verlag Berlin Heidelberg New York in 1991
Softcover reprint of the hardcover 1st edition 1991

Typesetting: Macmillan India Ltd, Bangalore-25;

51/3020-5 4 3 2 1 0 – Printed on acid-free paper

Editorial Board

Table of Contents

Iron-Containing Proteins and Related Analogs – Complementary Mössbauer, EPR and Magnetic Susceptibility Studies

Alfred X. Trautwein, Eckhard Bill, Emile L. Bominaar and Heiner Winkler

Institut für Physik, Medizinische Universität, Ratzeburger Allee 160, 2400 Lübeck, FRG

The three methods covered by this review – Mössbauer spectroscopy, electron paramagnetic resonance and magnetic susceptibility – provide a powerful set of tools for detailed studies of electronic structure and molecular magnetism of iron-containing proteins and related analogs. The interpretation of measured data is based on the spin-Hamiltonian concept which is described in detail. A short introduction to the principles of the three methods as well as a basic description of the spectrometers is given. The major part of the review deals with applications, e.g. to mononuclear iron complexes and to spin-coupled iron complexes. Wherever possible, the complementarity in applying the three methods is described.

Structure and Bonding 78
© Springer-Verlag Berlin Heidelberg 1991

1 Introduction

Iron-containing proteins have attracted the special attention of spectroscopists active in the field of biologically relevant molecules, because they allow the application of Mössbauer spectroscopy in combination with the other techniques. For the elucidation of structure–function relationships in metalloproteins structural and spectroscopic techniques have both been essential, e.g., in deciphering the geometry at the metal ion center, in describing the nature of the bonding between ligands and metal, and in determining the energies of excited states. In the pursuance of this program much emphasis is laid also on possible relations between the properties of metalloenzymes and biomimetic analogs. It has been delineated in a number of review articles that Mössbauer spectroscopy has played a major role in this achievement, at least as far as iron complexes are concerned [1–5].

Mössbauer spectroscopy has among others the advantage that without application of a magnetic field already a large amount of valuable information about spin and valence states as well as site symmetry can be extracted from the quadrupole splitting and the isomer shift. The full power of the method, however, is only obtained if field- and temperature-dependent measurements are performed. On the other hand, it is apparent that under these circumstances the information about the electron configuration is indirect because the nuclear environment is only sensed by the Mössbauer nucleus through mediation of the hyperfine coupling. This is due to the fact that the magnetic splittings observed in a Mössbauer spectrum are caused by a magnetic hyperfine field which is given as the product of magnetic hyperfine tensor and spin expectation values of the electrons. Therefore, much of the relevant chemical information, related to zero-field splitting, electron spin coupling and spin fluctuations, only becomes accessible when the magnetic hyperfine tensor is known. It is this point where the need for complementary and independent data has its origin.

Combined together, the three methods covered by this review, i.e., Mössbauer spectroscopy, electron paramagnetic resonance (EPR) spectroscopy and magnetic susceptibility, provide a powerful set of tools for detailed studies of the various aspects of molecular magnetism. The question, not finally answered up till now, is whether this molecular magnetism in itself contributes in any way to the function of the proteins. The answer is likely to be negative, because large magnetic fields are not found in living nature, and where organisms are exposed to them artificially for short times they seem to have little or no effect. However, the magnetic state of metals in biological systems is a key chemical parameter, especially when it comes to spin coupling. The same overlap of electronic orbitals which is crucial for spin coupling also controls, e.g., electron transfer processes. Systematic studies of molecular magnetism have something in common with many other areas of spectroscopy insofar as they serve two objectives, on the one hand, to learn how to deduce the spectroscopic properties of a system theoretically from its structure by going back as far as possible to first principles of quantum mechanics and Coulomb interactions, and, on the other hand, to draw conclusions about the geometric conformation and electronic configuration from spectroscopic findings.

The above-mentioned systematics which is needed can, of course, not be brought forward by studying native and mutated proteins alone, because the possibilities for manipulation are limited. Therefore, the efforts of bioinorganic chemistry to provide suitable model compounds are indispensable and a large part of this review will deal with studies of the latter.

2 The Spin-Hamiltonian Concept

The quantitative behavior of paramagnetic complexes under the influence of an external magnetic field and as a function of temperature can, with rare exceptions, be described adequately in the framework of the spin-Hamiltonian parametrization, which was introduced originally into paramagnetic resonance spectroscopy by Abragam and Pryce [6].

This method is applicable as long as ligand field and spin–orbit interactions produce a group of ground state atomic levels whose internal splitting is small compared to the energy separation from all higher levels, as it is nearly always the case in coordination complexes of transition metals. At sufficiently low temperatures, when only this low-lying group of levels is appreciably populated, it alone determines the properties of the system. In this case the spin-Hamiltonian method derives a fictitious spin quantum number S within this set of levels by attributing an empirical multiplicity $2S + 1$ to it. In principle, there is no necessity that this S, so defined, should be equal to the free ion spin, although in high-spin iron complexes this is in general the case. Beyond that, it is customary in EPR work to attribute an effective spin $S^{eff} = 1/2$ to the resonantly absorbing level pair, irrespective of its electronic character.

The Hamiltonian, which is finally constructed to describe the influence of the ligands and of an applied magnetic field on the central metal ion, has to reflect the local symmetry of the environment. However, it has been found that the electronic ground state of a large class of cases is adequately described by a simple spin Hamiltonian, which takes into account only axial and rhombic terms, i.e.,

$$\mathscr{H}_e = D[S_z^2 - S(S + 1)/3 + (E/D)(S_x^2 - S_y^2)] + \beta \vec{B}^{ext} \cdot \tilde{g} \cdot \vec{S}, \tag{1}$$

where D and E/D are the zero-field splitting and the rhombicity parameter, respectively, $\vec{B}^{ext}$ is the applied (external) field, $\tilde{g}$ the electronic g tensor and β the Bohr magneton. As will be shown in Sect. 2.3 the value of the rhombicity parameter can be restricted to the range $0 \leq E/D \leq 1/3$ by a suitable choice of the coordinate axes.

Even in the absence of an external field the $(2S + 1)$-fold degeneracy of the ground state is partly removed by the zero-field splitting, which has the form of an electrostatic quadrupole interaction and is derived from the admixing of excited orbital states in second order by spin–orbit interaction. When rhombic distortion is present any degeneracy in integer-spin systems is totally lifted but at least a twofold degeneracy is left in half-integer-spin systems as a consequence of Kramers' theorem. Kramers degeneracy will, however, be removed by the action of the Zeeman interaction $\beta \vec{B}^{ext} \cdot \tilde{g} \cdot \vec{S}$. In tetragonal symmetry the eigenstates of the spin Hamiltonian (1) can be labeled by the magnetic quantum number M_s as long as the external field is directed along the z axis. In the general case $(E/D \neq 0)$ the eigenvalues ε_j and the eigenstates $|\phi_j\rangle, j = 1, \ldots, (2S + 1)$, are mixtures of states with different M_s. By diagonalizing the matrix of the spin Hamiltonian $\mathscr{H}_e$ one obtains eigenvectors

$$|\phi_j\rangle = \sum_{M_s} C^j_{M_s} |S, M_s\rangle, \tag{2}$$

where $C^j_{M_s}$ are the expansion coefficients of $|\phi_j\rangle$ in the basis $|S, M_s\rangle$. Spin expectation values $\langle S_i \rangle_j \equiv \langle \phi_j | S_i | \phi_j \rangle$, $i = x, y, z$, may be calculated for each individual eigenstate:

$$\langle S_i \rangle_j = \sum_{M_s M_s'} C^j_{M_s} C^{j*}_{M_s'} \langle S, M_s' | S_i | S, M_s \rangle. \tag{3}$$

These $\langle S_i \rangle_j$ values generate internal fields at the nucleus of the paramagnetic center leading to hyperfine interactions. The magnitudes and the directions of the internal fields depend on the parameters D, E/D and $\vec{B}^{ext}$. It will be shown that the spin expectation values of a Kramers system are usually saturated by applied fields much faster than those of a non-Kramers system.

In the following, we will discuss in more detail the effect of the zero-field parameters D, E/D and of the external field $\vec{B}^{ext}$ on the eigenstates of the Hamiltonian (1) and on the spin expectation values in two examples, i.e. a Kramers system, $S = 5/2$, and a non-Kramers systems, $S = 2$.

2.1 The Spin Hamiltonian for High-Spin Ferric Iron $(S = 5/2)$

The Spin Hamiltonian for high-spin ferric iron is given by formula (1) using $S = 5/2$. Written down in the $|5/2, M_s\rangle$ representation it yields the matrix

M_s	$+5/2$	$+3/2$	$+1/2$	$-1/2$	$-3/2$	$-5/2$
$+5/2$	$\frac{10}{3}D + \frac{5}{2}H_z$	$\frac{\sqrt{5}}{2}(H_x - iH_y)$	$\sqrt{10}E$			
$+3/2$	$\frac{\sqrt{5}}{2}(H_x + iH_y)$	$-\frac{2}{3}D + \frac{3}{2}H_z$	$\sqrt{2}(H_x - iH_y)$	$3\sqrt{2}E$		
$+1/2$	$\sqrt{10}E$	$\sqrt{2}(H_x + iH_y)$	$-\frac{8}{3}D + \frac{1}{2}H_z$	$\frac{3}{2}(H_x - iH_y)$	$3\sqrt{2}E$	
$-1/2$		$3\sqrt{2}E$	$\frac{3}{2}(H_x + iH_y)$	$-\frac{8}{3}D - \frac{1}{2}H_z$	$\sqrt{2}(H_x - iH_y)$	$\sqrt{10}E$
$-3/2$			$3\sqrt{2}E$	$\sqrt{2}(H_x + iH_y)$	$-\frac{2}{3}D - \frac{3}{2}H_z$	$\frac{\sqrt{5}}{2}(H_x - iH_y)$
				$\sqrt{10}E$	$\frac{\sqrt{5}}{2}(H_x + iH_y)$	$\frac{10}{3}D - \frac{5}{2}H_z$

$$\tag{4}$$

where H_i denotes the term $\beta g_i B_i^{ext}$, $i = x, y, z$. To discern the influence of each individual parameter the following three cases are considered, where for convenience $g_x = g_y = g_z = 2$ has been taken.

2.1.1 Zero-Field Interaction Dominates Zeeman Interaction

When zero-field interaction dominates, i.e. $|D| \gg |\beta g B^{ext}|$, the terms $\beta g_i B_i^{ext}$, $i = x, y, z$, may be neglected in the matrix (4) yielding the following eigenvalues

ε_j and eigenstates $|\phi_j\rangle$:

$$\varepsilon_1 = -2D\alpha\cos[(\pi - \arccos\beta)/3], \tag{5a}$$

$$|\phi_1^{\pm}\rangle = \frac{1}{\left[1 + \dfrac{18E^2}{(\frac{2}{3}D + \varepsilon_1)^2} + \dfrac{10E^2}{(\frac{10}{3}D - \varepsilon_1)^2}\right]^{1/2}}$$

$$\times \left[|\pm 1/2\rangle + \frac{3\sqrt{2}E}{\frac{2}{3}D + \varepsilon_1}|\mp 3/2\rangle - \frac{\sqrt{10}E}{\frac{10}{3}D - \varepsilon_1}|\pm 5/2\rangle\right],$$

$$\varepsilon_2 = -2D\alpha\cos[(\pi + \arccos\beta)/3], \tag{5b}$$

$$|\phi_2^{\pm}\rangle = \frac{1}{\left[1 + \dfrac{(\frac{2}{3}D + \varepsilon_2)^2}{18E^2} + \dfrac{5(\frac{2}{3}D + \varepsilon_2)^2}{9(\frac{10}{3}D - \varepsilon_2)^2}\right]^{1/2}}$$

$$\times \left[|\mp 3/2\rangle + \frac{\frac{2}{3}D + \varepsilon_2}{3\sqrt{2}E}|\pm 1/2\rangle - \frac{\sqrt{\frac{5}{9}}(\frac{2}{3}D + \varepsilon_2)}{\frac{10}{3}D - \varepsilon_2}|\pm 5/2\rangle\right],$$

$$\varepsilon_3 = 2D\alpha\cos[(\arccos\beta)/3], \tag{5c}$$

$$|\phi_3^{\pm}\rangle = \frac{1}{\left[1 + \dfrac{(\frac{10}{3}D - \varepsilon_3)^2}{10E^2} + \dfrac{9(\frac{10}{3}D - \varepsilon_3)^2}{5(\frac{2}{3}D + \varepsilon_3)^2}\right]^{1/2}}$$

$$\times \left[|\pm 5/2\rangle - \frac{\frac{10}{3}D - \varepsilon_3}{\sqrt{10}E}|\pm 1/2\rangle - \frac{\sqrt{\frac{9}{5}}(\frac{10}{3}D - \varepsilon_3)}{\frac{2}{3}D + \varepsilon_3}|\mp 3/2\rangle\right],$$

where $\quad \alpha = [28(1 + 3E^2/D^2)/9]^{1/2}, \; \beta = 80(1 - 9E^2/D^2)/(3\alpha)^3. \tag{6}$

Without an applied field the eigenstates are twofold degenerate in the whole range $0 \leq E/D \leq 1/3$, according to Kramers' theorem. In Figs. 1 and 2 the eigenvalues and the spin expectation values $\langle S_i\rangle$, i = x, y, z, respectively, are plotted as a function of E/D in a weak external field (≈ 20 mT). The latter is required to remove the Kramers degeneracy and to define a quantization axis. In the following, two special cases are discussed.

(i) $E/D = 0$
When the rhombicity E/D is zero the expressions (5a–c) reduce to:

$$|\phi_1^{\pm}\rangle = |5/2, \pm 1/2\rangle \quad \text{for } \varepsilon_1 = -8D/3, \tag{7a}$$

$$|\phi_2^{\pm}\rangle = |5/2, \mp 3/2\rangle \quad \text{for } \varepsilon_2 = -2D/3, \tag{7b}$$

$$|\phi_3^{\pm}\rangle = |5/2, \pm 5/2\rangle \quad \text{for } \varepsilon_3 = +10D/3. \tag{7c}$$

Application of a weak field in a direction given by Θ, ϕ will remove the degeneracy and lead to states $|M_s\rangle$ with $M_s = \pm 5/2$ and $\pm 3/2$, and $e^{\pm i\Phi/2}\cos\frac{1}{2}\Theta|+M_s\rangle \pm e^{\pm i\Phi/2}\sin\frac{1}{2}\Theta|-M_s\rangle$ with $M_s = \pm 1/2$ as eigenvectors.

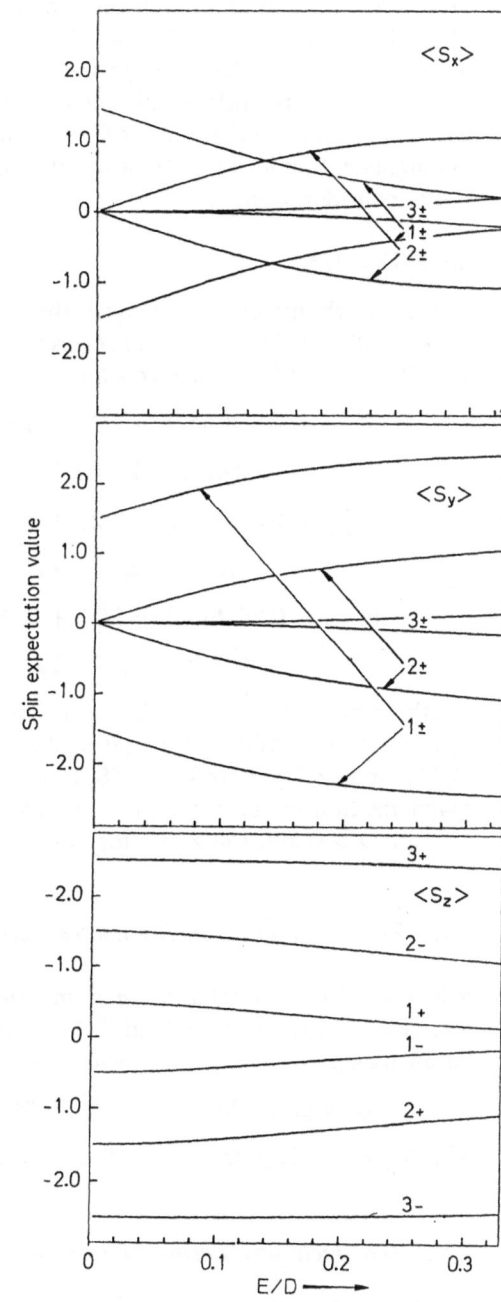

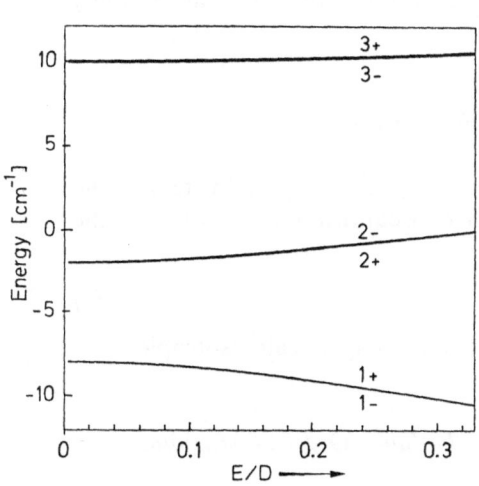

Fig. 1. Eigenvalues of the spin Hamiltonian (1) for $S = 5/2$ calculated as a function of E/D, with $D = 3\ cm^{-1}$ and $\vec{B}^{ext}(|\vec{B}^{ext}| = 20\ mT)$ pointing in z direction

Fig. 2. Spin expectation values $\langle S_i \rangle$, $i = x,\ y,\ z$, for $S = 5/2$ calculated from the spin Hamiltonian (1) as a function of E/D, with $D = 3\ cm^{-1}$ and $\vec{B}^{ext}$ ($|\vec{B}^{ext}| = 20\ mT$) pointing in x, y, z direction, respectively

Hence, as spin expectation values one obtains $|\langle S_x \rangle| = |\langle S_y \rangle| = 0$ and $\langle S_z \rangle = \pm 5/2$ for $M_s = \pm 5/2$, $|\langle S_x \rangle| = |\langle S_y \rangle| = 0$ and $\langle S_z \rangle = \pm 3/2$ for $M_s = \pm 3/2$, as well as $|\langle S_x \rangle| \approx |\langle S_y \rangle| \approx 3/2$ and $\langle S_z \rangle \approx \pm 1/2$ for $M_s = \pm 1/2$, respectively. When at low temperature only the lowest doublet is considered, there exists an easy plane of magnetization for $D > 0$ (states with $M_s = \pm 1/2$) and an easy axis of magnetization for $D < 0$ (states with $M_s = \pm 5/2$) in the paramagnetic system under consideration.

(ii) $E/D = 1/3$

When the rhombicity E/D takes the value $1/3$, the eigenvectors are linear combinations of M_s-states. The corresponding eigenvalues become: $\varepsilon_1 \approx -3.5D$, $\varepsilon_2 = 0$, $\varepsilon_3 \approx +3.5D$. The eigenstates are:

$$|\phi_1^\pm\rangle = 0.8882|5/2, \pm 1/2\rangle - 0.1363|5/2, \pm 5/2\rangle$$
$$- 0.4388|5/2, \mp 3/2\rangle, \tag{8a}$$

$$|\phi_2^\pm\rangle = 0.8965|5/2, \mp 3/2\rangle + 0.4224|5/2, \pm 1/2\rangle$$
$$- 0.1334|5/2, \pm 5/2\rangle, \tag{8b}$$

$$|\phi_3^\pm\rangle = 0.9816|5/2, \pm 5/2\rangle + 0.1808|5/2, \pm 1/2\rangle$$
$$+ 0.0609|5/2, \mp 3/2\rangle, \tag{8c}$$

Furthermore, one obtains the spin expectation values with the feature that $|\langle S_y \rangle| \gg |\langle S_x \rangle|$, $|\langle S_z \rangle|$ for states $|\phi_1^\pm\rangle$, $|\langle S_x \rangle| \approx |\langle S_y \rangle| \approx |\langle S_z \rangle|$ for states $|\phi_2^\pm\rangle$, and $|\langle S_z \rangle| \gg |\langle S_x \rangle|$, $|\langle S_y \rangle|$ for states $|\phi_3^\pm\rangle$. Thus an easy axis of magnetization exists in the ground state doublet for both cases, namely the y axis for $D > 0$ and the z axis for $D < 0$.

2.1.2 Zeeman Interaction Dominates Zero-Field Interaction

When the Zeeman interaction dominates, i.e. $|\beta g \vec{B}^{ext}| \gg |D|$, by taking the direction of the external field $\vec{B}^{ext}$ as axis of quantization one obtains the following eigenvalues and eigenvectors

$$\varepsilon_{M_s} \approx g\beta B^{ext} M_s, \quad |\phi_{M_s}\rangle \approx |S, M_s\rangle, \tag{9}$$

with $M_s = \pm 5/2, \pm 3/2, \pm 1/2$. The system is then magnetically isotropic.

2.1.3 Zero-Field and Zeeman Interaction of the Same Order of Magnitude

In the case that $|D| \approx |g\beta \vec{B}^{ext}|$, the eigenvalues and eigenvectors of the Hamiltonian (1) and consequently the spin expectation values are functions of all three parameters D, E/D and $\vec{B}^{ext}$. Therefore the values of D and E/D can conveniently be extracted from experimental data by magnetic measurements in

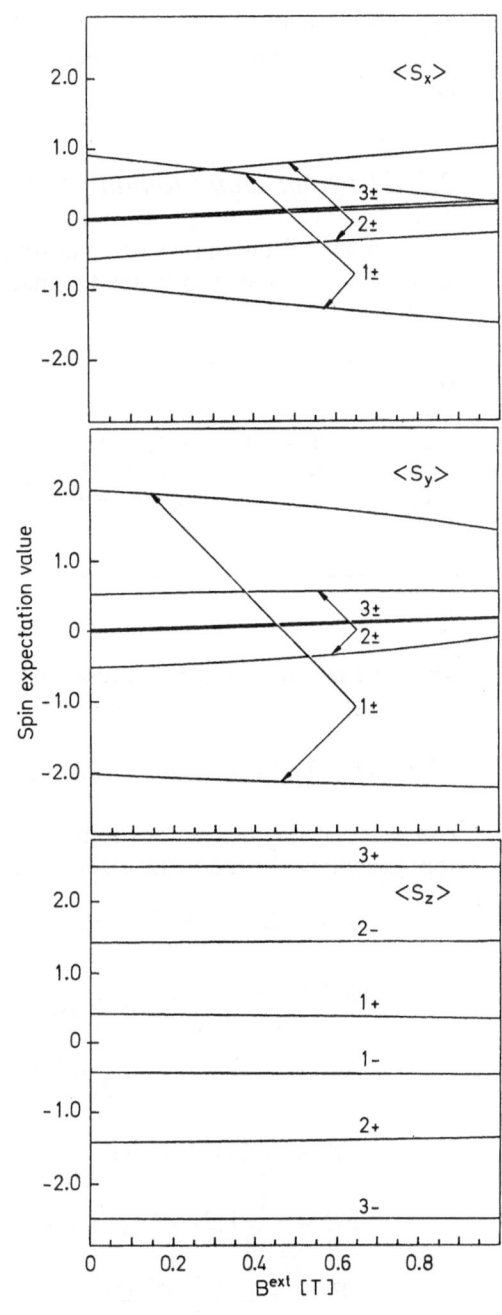

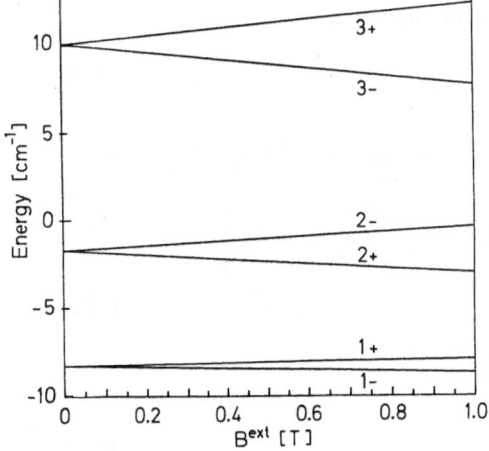

Fig. 3. Eigenvalues of the spin Hamiltonian (1) for $S = 5/2$ calculated as a function of $\vec{B}^{ext}$ pointing in z direction, with $D = 3\ cm^{-1}$ and $E/D = 0.1$

Fig. 4. Spin expectation values $\langle S_i \rangle$, $i = x, y, z$, for $S = 5/2$ calculated from the spin Hamiltonian (1) as a function of $\vec{B}^{ext}$ pointing in x, y, z direction, respectively, with $D = 3\ cm^{-1}$ and $E/D = 0.1$

appropriate fields. Figures 3 and 4 show the eigenvalues and the spin expectation values of spin S = 5/2, respectively, as functions of B^{ext} with D and E/D as given.

2.2 The Spin Hamiltonian for High-Spin Ferrous Iron (S = 2)

The spin Hamiltonian (1) written down in the $|2, M_s\rangle$ representation for a high-spin ferrous iron system yields the matrix:

M_s	+ 2	+ 1	0	− 1	− 2
+ 2	$2D + 2H_z$	$H_x - iH_y$	$\sqrt{6}E$		
+ 1	$H_x + iH_y$	$-D + H_z$	$\frac{1}{2}\sqrt{6}(H_x - iH_y)$	$3E$	
0	$\sqrt{6}E$	$\frac{1}{2}\sqrt{6}(H_x + iH_y)$	$- 2D$	$\frac{1}{2}\sqrt{6}(H_x - iH_y)$	$\sqrt{6}E$ (10)
− 1		$3E$	$\frac{1}{2}\sqrt{6}(H_x + iH_y)$	$-D - H_z$	$H_x - iH_y$
− 2			$\sqrt{6}E$	$H_x + iH_y$	$2D - 2H_z$

where, as in matrix (4), H_i denotes the term $\beta g_i B_i^{ext}$, i = x, y, z.

2.2.1 Zero-Field Interaction Dominates Zeeman Interaction

When zero-field interaction dominates the Zeeman interaction, one obtains approximately the following eigenvalues and eigenvectors:

$$\varepsilon_1 = -2D\sqrt{1 + 3E^2/D^2}, \tag{11a}$$

$$|\phi_1\rangle = \frac{1}{[(1 + \sqrt{1 + 3E^2/D^2})^2 + 3E^2/D^2]^{1/2}} \left[-\frac{\sqrt{3}}{2}\frac{E}{D}|2, -2\rangle \right.$$
$$\left. + (1 + \sqrt{1 + 3E^2/D^2})|2, 0\rangle - \frac{\sqrt{3}}{2}\frac{E}{D}|2, +2\rangle \right];$$

$$\varepsilon_2 = -D(1 + 3E/D), \tag{11b}$$

$$|\phi_2\rangle = \frac{1}{\sqrt{2}}[|2, -1\rangle - |2, +1\rangle];$$

$$E_3 = -D(1 - 3E/D), \tag{11c}$$

$$|\phi_3\rangle = \frac{1}{\sqrt{2}}[|2, -1\rangle + |2, +1\rangle];$$

$$\varepsilon_4 = +2D, \tag{11d}$$

$$|\phi_4\rangle = \frac{1}{\sqrt{2}}[|2, -2\rangle - |2, +2\rangle];$$

$$\varepsilon_5 = +2D\sqrt{1 + 3E^2/D^2}, \tag{11e}$$

$$|\phi_5\rangle = \frac{1}{[(1 + \sqrt{1 + 3E^2/D^2})^2 + 3E^2/D^2]^{1/2}}$$

$$\times\left[\frac{1}{\sqrt{2}}(1 + \sqrt{1 + 3E^2/D^2})|2, -2\rangle\right.$$

$$\left. + \sqrt{3}\frac{E}{D}|2, 0\rangle + \frac{1}{\sqrt{2}}(1 + \sqrt{1 + 3E^2/D^2})|2, +2\rangle\right].$$

In the same way as for high-spin ferric iron ($S = 5/2$) the sign of D determines the lowest eigenstate; in the present case $|\phi_1\rangle$ for $D > 0$ and $|\phi_5\rangle$ for $D < 0$. One observes that all eigenstates for $S = 2$ are singlets as long as $E/D \neq 0$; this situation is different from that for a Kramers system. The mixing of states with $M_s = 0, \pm 2$ in $|\phi_1\rangle$ and $|\phi_5\rangle$ increases as E/D departs from zero, while the states with $M_s = \pm 1$ remain unaffected. In the following it will be shown that the value of E/D influences the saturation behavior of spin expectation values in the applied fields.

(i) $E/D = 0$

For $E/D = 0$ one obtains the eigenvalues and eigenstates as follows:

$$|\phi_1\rangle = |2, 0\rangle \qquad \text{for} \quad \varepsilon_1 = -2D, \tag{12a}$$

$$|\phi_{2,3}\rangle = |2, \pm 1\rangle \qquad \text{for} \quad \varepsilon_{2,3} = -D, \tag{12b}$$

$$|\phi_{4,5}\rangle = |2, \pm 2\rangle \qquad \text{for} \quad \varepsilon_{4,5} = +2D. \tag{12c}$$

As shown in Figs. 5 and 6, respectively, application of an external field generates spin expectation values with the feature that $|\langle S_x\rangle|, |\langle S_y\rangle| \gg |\langle S_z\rangle|$ for $|\phi_1\rangle$ and $|\langle S_z\rangle| \gg |\langle S_x\rangle|, |\langle S_y\rangle|$ for $|\phi_{4,5}\rangle$, and thus an easy plane of magnetization exists for $D > 0$ and an easy axis of magnetization for $D < 0$.

(ii) $E/D = 1/3$

For $E/D = 1/3$ one obtains

$$|\phi_1\rangle = 0.9659|2, 1\rangle - 0.1830|2, +2\rangle - 0.1830|2, -2\rangle \quad \text{for} \quad \varepsilon_1 = -4D/\sqrt{3}, \tag{13a}$$

$$|\phi_2\rangle = 0.707|2, -1\rangle - 0.707|2, +1\rangle \qquad\qquad \text{for} \quad \varepsilon_2 = -2D, \tag{13b}$$

$$|\phi_3\rangle = 0.707|2, -1\rangle + 0.707|2, +1\rangle \qquad\qquad \text{for} \quad \varepsilon_3 = 0, \tag{13c}$$

$$|\phi_4\rangle = 0.707|2, +2\rangle - 0.707|2, -2\rangle \qquad\qquad \text{for} \quad \varepsilon_4 = +2D, \tag{13d}$$

$$|\phi_5\rangle = 0.6832|2, +2\rangle + 0.6832|2, -2\rangle + 0.2588|2, 0\rangle \quad \text{for} \quad \varepsilon_5 = +4D/\sqrt{3}. \tag{13e}$$

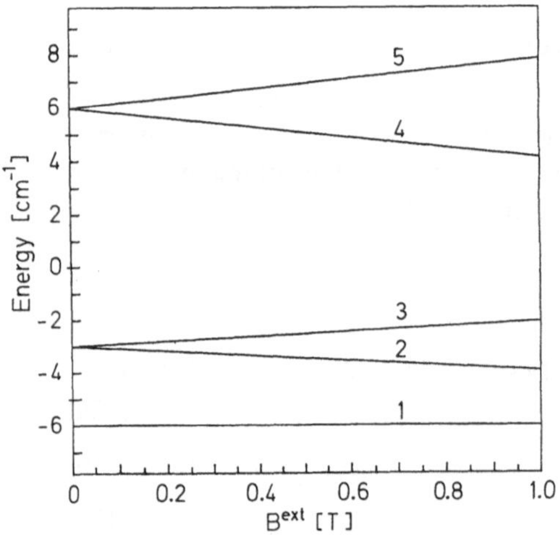

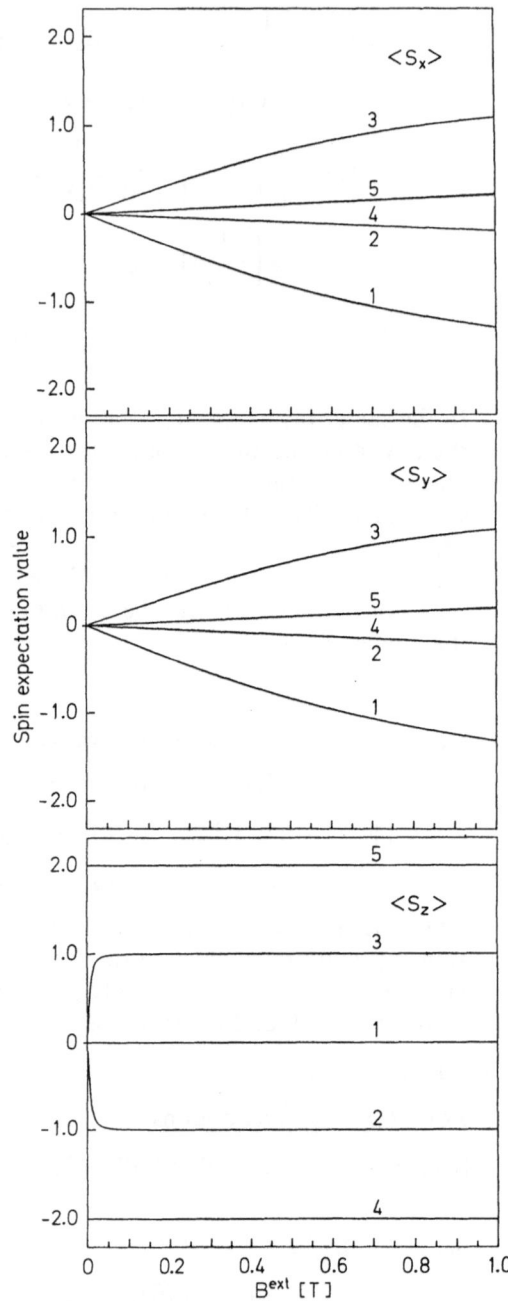

Fig. 5. Eigenvalues of the spin Hamiltonian (1) for $S = 2$ calculated as a function of $\vec{B}^{ext}$ pointing in z direction, with $D = 3\ cm^{-1}$ and $E/D = 0$

Fig. 6. Spin expectation values $\langle S_i \rangle$, $i = x, y, z$, for $S = 2$ calculated from the spin Hamiltonian (1) as a function of $\vec{B}^{ext}$ pointing in x, y, z direction, with $D = 3\ cm^{-1}$ and $E/D = 0$

Spin expectation values with the feature that $|\langle S_y \rangle| \gg |\langle S_x \rangle|, |\langle S_z \rangle|$ for $|\phi_{1,2}\rangle$, $|\langle S_x \rangle| \approx |\langle S_y \rangle| \approx |\langle S_z \rangle|$ for $|\phi_3\rangle$ and $|\langle S_z \rangle| \gg |\langle S_x \rangle|, |\langle S_y \rangle|$ for $|\phi_{4,5}\rangle$ are found, as shown in Figs. 7 and 8. Thus an easy axis of magnetization exists for both $D < 0$ ($|\phi_5\rangle$ lies lowest) and $D > 0$ ($|\phi_1\rangle$ lies lowest).

Figures 9 and 10 give the eigenvalues and spin expectation values of Eq. (1) for $S = 2$, respectively, as functions of B^{ext} calculated with $D = 3 \text{ cm}^{-1}$, $E/D = 0.1$. By comparing Fig. 10 with Figs. 6 and 8, where the value of E/D is 0.1, 0, and 0.33, respectively, one observes that the value of E/D affects the saturation behavior of spin expectation values.

2.2.2 Zeeman Interaction Dominates Zero-Field Interaction

When the Zeeman interaction dominates the zero-field interaction one gets equivalent eigenvalues and eigenvectors as given in Sect. 2.1.2 by taking the direction of the applied field as axis of quantization, i.e. the expressions (9), however with $M_s = -2, -1, \ldots, +2$. The system is again magnetically isotropic.

2.2.3 Zero-Field and Zeeman Interaction of the Same Order of Magnitude

Under the condition that zero-field and Zeeman interaction are of the same order of magnitude, the eigenvalues and eigenvectors, as well as the spin expectation values, are functions of all three parameters D, E/D and $\vec{B}^{ext}$. By comparing Figs. 9, 10 with Figs. 3, 4 one notices that the spin expectation values of a non-Kramers system exhibit a gradual approach to the saturation value in contrast to a Kramers system, which is saturated even in very small fields.

2.3 The Theoretical Background of the Spin-Hamiltonian Formalism

The electronic structure of atomic and molecular aggregates is mainly determined by the mutual Coulomb interactions of electrons and nuclei in these systems. These interactions lead in atoms to the well-known energy terms of the form ^{2S+1}X, with $X = S, P, D, \ldots$ representing the angular-momentum quantum numbers $L = 0, 1, 2, \ldots$, and $2S + 1$ the spin multiplicity (degeneracy). According to Hund's rule, the electronic ground term in a free iron(II) ion is ^5D, which possesses a five-fold spin and orbital degeneracy. In molecular systems, however, the orbital degeneracy of the high-spin iron(II) ion is usually lifted as a consequence of the 'ligand-field' interaction between the 3d-electrons and the ligands. The iron(III) ion has already in absence of a ligand field an orbitally non-degenerate ground term, ^6S. The spin degeneracy is not affected by Coulomb interactions. The zero-field splittings in the spin multiplets are mainly

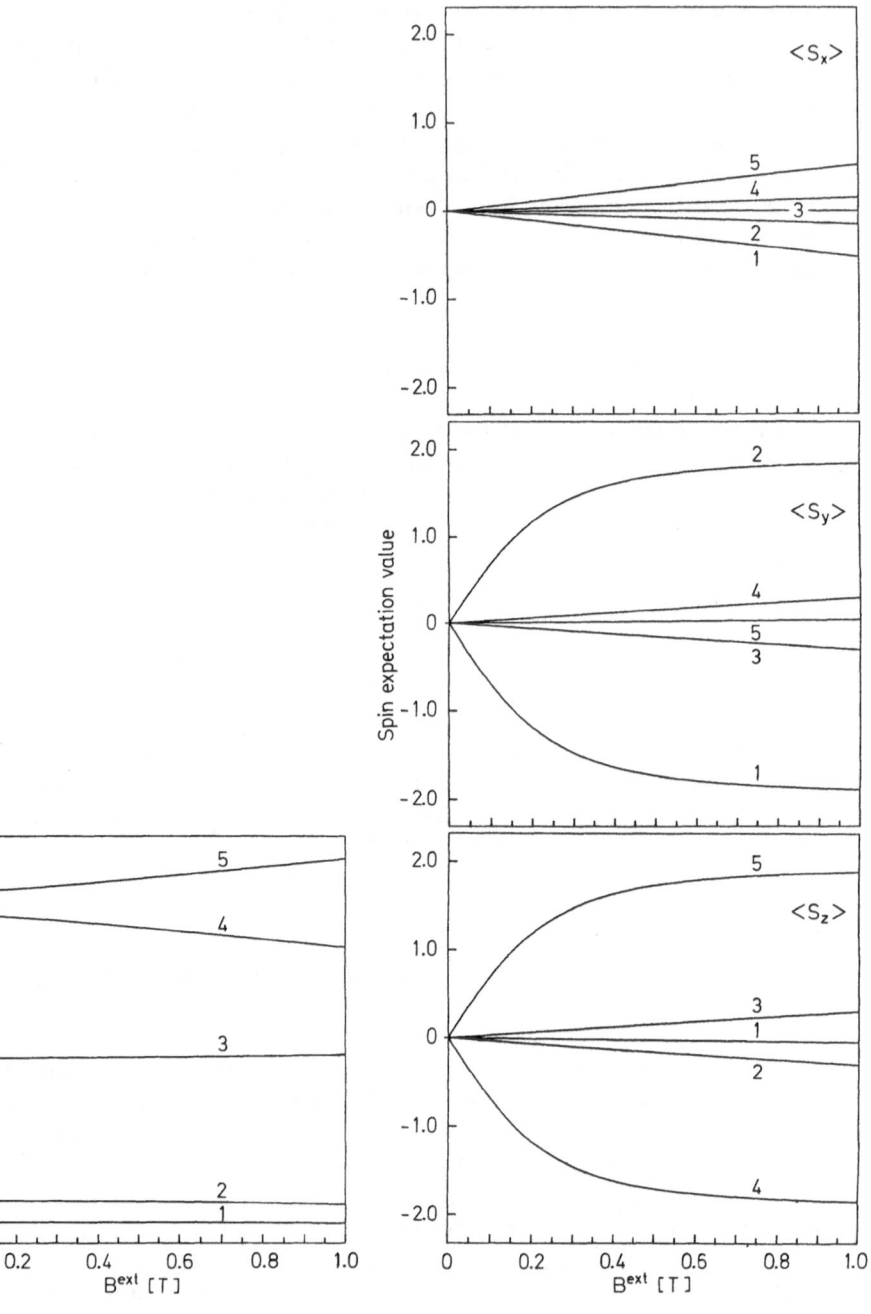

Fig. 7. Eigenvalues of the spin Hamiltonian (1) for S = 2 calculated as a function of $\vec{B}^{ext}$ pointing in z direction, with D = 3 cm^{-1} and E/D = 0.33

Fig. 8. Spin expectation values $\langle S_i \rangle$, i = x, y, z, for S = 2 calculated from the spin Hamiltonian (1) as a function of $\vec{B}^{ext}$ pointing in x, y, z direction, respectively, with D = 3 cm^{-1} and E/D = 0.33

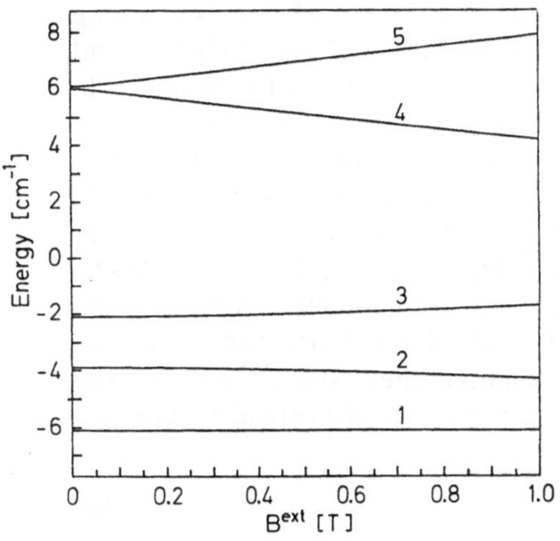

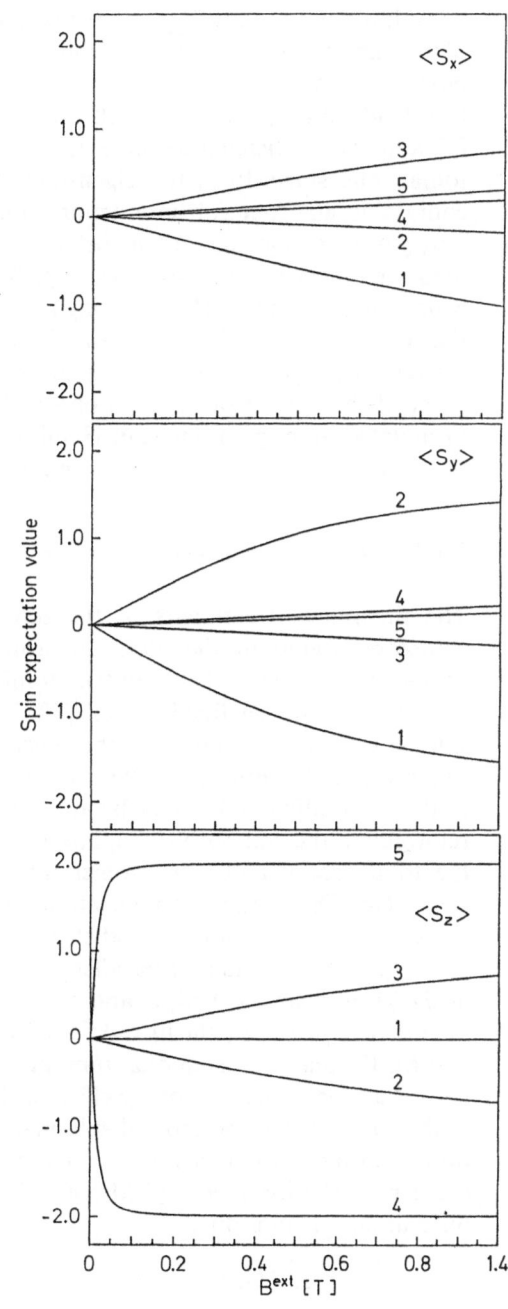

Fig. 9. Eigenvalues of the spin Hamiltonian (1) for S = 2 calculated as a function of $\vec{B}^{ext}$ pointing in z direction, with D = 3 cm^{-1} and E/D = 0.1

Fig. 10. Spin expectation values $\langle S_i \rangle$, i = x, y, z, for S = 2 calculated from the spin Hamiltonian (1) as a function of $\vec{B}^{ext}$ pointing in x, y, z direction, respectively, with D = 3 cm^{-1} and E/D = 0.1

a result of the coupling $\zeta\sum_i \vec{l}_i \cdot \vec{s}_i$ between the spins $\vec{s}_i$ and the orbital momenta $\vec{l}_i$ of the unpaired electrons (spin–orbit coupling). For completeness we notice that next to the spin–orbit interactions also spin–spin interactions contribute to the zero-field splittings. However, the latter contributions are small (< 0.5 cm^{-1}) [7] and will be disregarded here. In the derivation of the fine-structure Hamiltonian one starts from the eigenstates $|n\rangle$ of the electrostatic Hamiltonian containing all Coulomb interactions, including the ligand field. We consider only paramagnetic ions with an orbitally non-degenerate ground state, $|0\rangle$, well-separated from all excited states, $|n\rangle$. By this we mean that any excited state which mixes by spin–orbit coupling into the ground state lies at energy E_n such that $E_n - E_0 \gg |\zeta|$. This condition allows us to treat the spin–orbit interaction between the ground and excited states in second-order perturbation theory. The effect of the spin–orbit interaction on the ground spin manifold can then be formulated in terms of the spin Hamiltonian given in Eq. (1) in both the high-spin iron(II) and iron(III) complexes, as outlined in the following section.

2.3.1 Iron (II) Complexes

The theoretical treatments of the zero-field splittings in high-spin iron(II) complexes found in the literature usually start from the premise that the spin–orbit interaction between the atomic ground term (^{5}D) and excited terms (^{3}P, ^{3}D, . . .) are negligibly small [7]. The zero-field splittings are then considered to result entirely from the effect of the spin–orbit interaction within a single atomic LS term, notably the term ^{5}D in high-spin iron(II). The restriction to the ^{5}D multiplet allows us, by application of the Wigner–Eckart theorem, to reformulate the spin–orbit coupling operator as $\lambda\vec{L}\cdot\vec{S}$, with $\vec{L}$ and $\vec{S}$ denoting the total orbital and spin momentum, respectively, of the iron(II) ion, and $\lambda = -\zeta/4$. The ground configuration 3d^6 of high-spin iron(II) in ligand field corresponds with a half-filled 3d shell containing five spin-parallel 'α' electrons plus a 'β' electron with anti-parallel spin. The orbitals occupied by the β electron in the ground state ($|0, M_s\rangle$) and excited states ($|n, M_s\rangle$, $n > 0$) of the iron(II) ion are determined by the ligand field. In the notation, n labels the orbital state and M_s the magnetic spin quantum number, $M_s = -S, -S + 1, \ldots, S$. In the perturbational treatment of the effect of the spin–orbit coupling on the energies of the sub-states in the ground spin manifold $|0, M_s\rangle$, the first-order contributions vanish due to quenching of the orbital momentum: $\langle 0|\vec{L}|0\rangle = 0$. Second-order perturbation theory yields, after integration over the orbital variables, a bilinear spin Hamiltonian

$$\mathscr{H} = \sum_{p,q} D_{pq} S_p S_q \tag{14a}$$

describing the zero-field splitting in the ground state manifold, with the fine-structure tensor given by

$$D_{pq} = -\lambda^2 \sum_n{}' \frac{\langle 0|L_p|n\rangle\langle n|L_q|0\rangle}{E_n - E_0}. \tag{14b}$$

Third- and higher-order corrections are smaller than D_{pq} by factors of the order $|\lambda|/(E_n - E_0) \ll 1$ and are disregarded. After transformation of D_{pq} to principal axes $(\hat{X}, \hat{Y}, \hat{Z})$ and removal of the trace, the fine-structure tensor is cast in the diagonal form $d_{\hat{x}}S_{\hat{x}}^2 + D_{\hat{y}}S_{\hat{y}}^2 + d_{\hat{z}}S_{\hat{z}}^2$, with $d_{\hat{x}} + d_{\hat{y}} + d_{\hat{z}} = 0$. The subtraction of the trace corresponds to a shift in the origin of the energy scale which is irrelevant because it does not affect the eigenstates and the differences between the eigenvalues. The Hamiltonian for the zero-field splitting can be expressed as

$$\mathscr{H} = D_z[S_z^2 - S(S + 1)/3 + E_z(S_x^2 - S_y^2)/D_z], \tag{15}$$

with $D_z = -(d_x + d_y - 2d_z)/2$ and $E_z = (d_x - d_y)/2$, for any permutation X, Y, Z of $\hat{X}, \hat{Y}, \hat{Z}$. In the definition of the zero-field parameters $D \equiv D_z$ and $E \equiv E_z$ in expression (1), we adopted the convention $|d_x| \le |d_y| \le |d_z|$ for which one can show readily that the inequality $0 \le E/D \le 1/3$ holds.

In order to illustrate the relation between the electronic structure and the zero-field parameters, we apply expressions (14a, b) to a high-spin iron(II) ion in a rhombically (C_{2v}) distorted octahedral ligand field. For simplicity, the summation in expression (14b) is restricted to the states in which the β electron occupies one of the t_{2g} orbitals, $|xy\rangle$, $|xz\rangle$, $|yz\rangle$. The zero-field parameters can then be expressed in terms of the ligand-field parameters Δ and w indicated in Fig. 11, as follows. In case the energy gap between the two lower levels is greater than or equal to the separation between the upper pair ($w/\Delta \le 1/3$) the zero-field parameters are $D = \lambda^2 \Delta/(\Delta^2 - w^2)$ and $E/D = w/\Delta$. D is then positive and the main principal axis of the D tensor is oriented perpendicular to the xy plane. When, however, the energy gap between the upper two orbital levels exceeds the separation between the lower pair ($w/\Delta > 1/3$), the zero-field parameters are $D = -\lambda^2(\Delta + 3w)/2(\Delta^2 - w^2)$ and $E/D = (\Delta - w)/(\Delta + 3w)$. D is then negative and the main axis of the D tensor is oriented in the xy plane. We note that the relative orientation of the principal axes of the D and EFG tensors can be established by Mössbauer analysis. In the example given in Fig. 11, the β electron occupies the $|xy\rangle$ orbital in the ground configuration. Admixtures of excited configurations into the ground state by spin–orbit coupling are of the

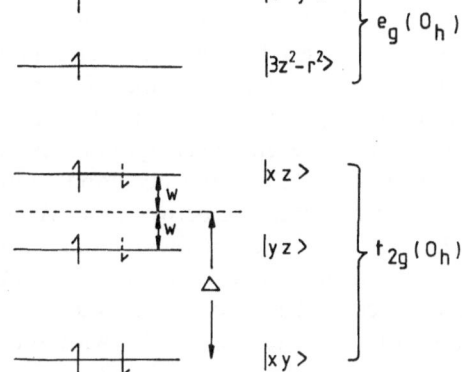

Fig. 11. One-electron orbital scheme of a high-spin (S = 2) iron(II) ion in a rhombic (C_{2v}) ligand field. *Full arrows* indicate orbital occupations and spin orientations in the 3d shell in the ground configuration. *Dashed arrows* indicate the orbitals occupied by the excess electron (↓) in the excited configurations considered in the text

order $|\lambda|/\Delta$ and are assumed to be small. Consequently, the main axis of the EFG tensor is oriented along the normal of the xy plane, i.e. for $w/\Delta \leqslant 1/3$ and $w/\Delta > 1/3$, parallel and perpendicular to the main axis of the D tensor, respectively. By extending the summation in Eq. (14b) to the e_g levels, the zero-field parameters can assume all possible orientations, signs of D, and values of E/D, depending on energy gaps between the e_g and t_{2g} orbitals [8].

In actual biomolecules, the ligand surrounding of the iron site may possess only C_1 symmetry. As a consequence, the orbital states become linear combinations of 3d orbitals, making the principal axes of the D, EFG, and A tensors noncollinear. The ligand-field calculations in Ref. [8] show that already slight deviations from tetragonal symmetry in the first coordination sphere of the iron ion may result in serious deviations from collinearity.

2.3.2 Iron(III) Complexes

The ground state in high-spin (S = 5/2) iron(III) complexes derives, as already noted, from the atomic ^6S term of the metal ion. Hence, all matrix elements of the spin–orbit coupling operator are identical zero within the ground manifold (L = 0). The vanishing of the spin–orbit interactions in the ground term of the iron(III) ion explains the minor zero-field splittings observed in the majority of the high-spin iron (III) complexes (D < 0.2 cm^{-1}), as compared with the iron(II) species ($|D| \sim 0$–15 cm^{-1}). Important exceptions are the high-spin iron(III) porphyrins which possess large zero-field parameters D ranging from 4 to 20 cm^{-1}. In order to interpret these large zero-field parameters, the ground term is assumed to interact through spin–orbit interaction with proximate ligand-field terms originating from excited atomic energy terms. A detailed ligand-field analysis of iron(III) complexes [9] reveals in octahedral ligand field the presence of a low-lying $^4T_1$ term, interacting with the ground term $^6A_1$ (i.e. ^6S in O_h symmetry) through spin–orbit coupling. In the evaluation of the interaction matrix it is essential to adopt the genuine spin–orbit coupling operator $\zeta \sum_i \vec{l}_i.\vec{s}_i$. In octahedral ligand field, the zero-field parameters D and E are identical zero for symmetry reasons. The $^4T_1$ term is then found to be uniformly admixed into the three Kramers doublets of the ground term $|^6A_i, M_s\rangle$, $M_s = \pm 1/2, \pm 3/2, \pm 5/2$, and the energy reductions of the three Kramers doublets, caused by spin–orbit coupling with excited ligand-field terms, are equal. However, due to the inequivalence of the equatorial and axial ligands in the iron porphyrins, the symmetry of the ligand field is lowered from O_h to tetragonal or even lower symmetries. Using the notation adequate to D_4 symmetry, the $^4T_1$ term is decomposed into $^4A_2$ and ^4E terms [9] (see Fig. 12). The energy separation between the $^4A_2$ and ^4E terms ($\Delta_2 - \Delta_1$, Fig. 12) causes different admixtures of these terms into the three Kramers doublets of the ground manifold, and leads thereby to non-vanishing zero-field splittings. The fine-structure Hamiltonian of the ground state, obtained by treating the spin–orbit interaction of $^6A_1$ with the $^4A_2$ and ^4E terms in second-order

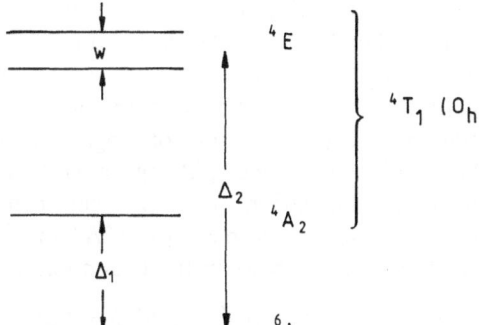

Fig. 12. Energy scheme of the many-electron terms of a high-spin $(S = 5/2)$ iron(III) ion in a ligand field of D_4 symmetry, considered in the derivation of the fine-structure operator presented in the text

perturbation theory, assumes the form DS_z^2, in which the zero-field parameter is given by

$$D = \frac{\zeta^2}{5}\left(\frac{1}{\Delta_1} - \frac{1}{\Delta_2}\right).$$ (16a)

Δ_1 and Δ_2 denote the energies of the $^4A_2$ and ^4E terms relative to the $^6A_1$ term. The positive sign of D found in high-spin iron(III) porphyrins indicates that the $^4A_2$ term is energetically below the ^4E term in these systems. Splitting of the ^4E term (w in Fig. 12), caused by a further lowering of the ligand-field symmetry, yields in second-order approximation and $w^2 \ll \Delta_2^2$, the rhombicity parameter

$$E/D = \frac{W\Delta_1}{2\Delta_2(\Delta_2 - \Delta_1)}$$ (16b)

The admixture of spin-quartet $(S = 3/2)$ terms into the spin sextet $(S = 5/2)$ ground state makes a 'high'-spin iron(III) complex actually a 'mixed'-spin complex. Nevertheless, the pure spin states of the manifold $|S = 5/2, M_s\rangle$ ought to be adopted in the spin-Hamiltonian analysis of the magnetic properties of slightly spin-admixed high-spin systems, as the effects of the spin mixture on the energies are put into the fine-structure Hamiltonian in Eq. (1). In this approach, the g values also need correction in order to accommodate the effect of the spin mixture on the Zeeman interaction. The expressions for the g values to be used in Eq. (1), obtained with first-order wavefunctions in tetragonal symmetry, read

$$g_{||} = 2 - \frac{8}{5}\left(\frac{\zeta}{\Delta_2}\right)^2$$ (17a)

$$g_\perp = 2 - \frac{4}{5}\left(\frac{\zeta}{\Delta_1}\right)^2 - \frac{4}{5}\left(\frac{\zeta}{\Delta_2}\right)^2.$$ (17b)

The expressions show that in octahedral symmetry $(\Delta_1 = \Delta_2)$, g is reduced isotropically below two by the admixture of the spin-quartet term. For $\Delta_2 > \Delta_1$, $g_\perp$ is more reduced than $g_{||}$.

The formulation of the fine-structure interactions as given in the spin Hamiltonian Eq. (1) loses its validity if the spin–orbit interaction cannot be adequately treated in second-order perturbation theory. This happens if the ligand-field splitting between the ground and excited terms, notably $^4A_2$, becomes of the same order-of-magnitude as the spin–orbit coupling constant ($\Delta_1 \sim \zeta$). The zero-field splittings must then be calculated by a non-perturbative, direct diagonalization of the interaction matrix. By restricting our considerations to the space spanned by the $^6A_1$ and $^4A_2$ manifolds, ($\Delta_2 = \infty$) the mixed-spin Kramers doublets in the ground state become

$$|\pm M_s\rangle = a_M|^6A_1, \pm M_s\rangle + b_M|^4A_2, \pm M_s\rangle, \tag{18}$$

for $M_s = 1/2$ and $3/2$. The $|\pm 5/2\rangle$ states are of pure high-spin character. The coefficients a_M and b_M in Eq. (18) are calculated by solving a 2×2 secular problem of the combined electrostatic and spin–orbit interaction. The diagonal elements of the interaction matrices are the electrostatic energies of the $^6A_1$ and $^4A_2$ terms; the off-diagonal elements, c_M, are due to the spin–orbit coupling operator, and are given by $c_{1/2} = -\zeta\sqrt{(6/5)}$ and $c_{3/2} = -\zeta\sqrt{(4/5)}$. The mixing coefficients a_M and b_M in Eq. (18) are a function of the ratio Δ_1/ζ. The relative energies of the Kramers doublets read

$$\varepsilon_M = (\Delta_1 - \sqrt{\Delta_1^2 + 4c_M^2})/2, \tag{19}$$

with $\varepsilon_{5/2} = 0$. For large term splittings, $|\Delta_1| \gg \zeta$, the zero-field splittings $\varepsilon_{5/2} - \varepsilon_{3/2}$ and $\varepsilon_{3/2} - \varepsilon_{1/2}$ approach the second-order results for these splittings, i.e. 4D and 2D, respectively, with D as given in Eq. (16a). When the term splitting goes from infinity to zero, the ratio of the two zero-field splittings increases from 2 to 4.45. In order to incorporate this increase into the spin-operator formulation of the zero-field splittings, one has to introduce an additional term into the spin ($S = 5/2$) Hamiltonian, e.g. of the form $D'S_z^4$.

The exact expressions for the g values appearing in the Zeeman term of a spin ($S = 5/2$)-Hamiltonian description of the mixed-spin $^6A_1 - ^4A_2$ state, read

$$g_{||} = 2 \tag{20a}$$

$$g_\perp = (6a_{1/2}^2 + 4b_{1/2}^2)/3. \tag{20b}$$

$g_\perp$ can be expressed in terms of Δ_1 and ζ, by substitution of the coefficients $a_{1/2}(\Delta_1, \zeta)$ and $b_{1/2}(\Delta_1, \zeta)$ obtained from diagonalization of the interaction matrices, as follows

$$g_\perp = [5 + 1/\sqrt{1 + 24(\zeta/\Delta_1)^2/5}]/3. \tag{20c}$$

Expression (20c) reduces to its second-order counterpart, Eq. (17b), for $\zeta/\Delta_1 \ll 1$ and $\Delta_2 = \infty$.

Concerning the magnetic hyperfine interactions in mixed-spin complexes, we note the following. The $^4A_2$ term possesses a non-spherical spin-distribution, which generates a non-vanishing spin-dipolar magnetic hyperfine field, in addition to the Fermi contact term. Since the amounts by which the $^4A_2$ term admixes into the three Kramers doublets of the ground state are different, the spin-dipolar contributions to the magnetic hyperfine field are doublet-dependent. The spin-dipolar hyperfine field can be incorporated into the spin (5/2)-Hamiltonian analysis by introducing doublet-dependent $\tilde{A}$ tensors in the calculation of the hyperfine fields: $\tilde{A}_M \cdot \langle \vec{S} \rangle_M$ [10].

In summary, we conclude that the spin $(S = 5/2)$-Hamiltonian formalism can be maintained in the analysis of strongly mixed-spin iron(III) complexes $(\Delta_1 \sim \zeta)$ provided a $D'S_z^4$ term is included in Eq. (1), the $g_\perp$ value is chosen according to Eq. (20c), and the $\tilde{A}$ tensors are allowed to depend on the Kramers doublet.

3 The Experimental Methods

The experimental methods will be sketched only briefly for reference in the subsequent discussions.

3.1 Mössbauer Spectroscopy

3.1.1 Principles

Mössbauer spectroscopy is a nuclear spectroscopic method. The term Mössbauer effect stands for recoil-free resonant absorption of gamma rays, which in the case of ^{57}Fe have an energy of 14.4 keV. Suppression of the energy loss, which would be caused by the recoil transmitted to the nuclei during emission or absorption of gamma quanta, is achieved by embedding the atoms in a solid matrix. The information about the electronic environment of the Mössbauer nucleus is therefore only accessible via the various hyperfine interactions between nuclei and electrons. The Hamiltonian which is adequate for the description of the Mössbauer nucleus ^{57}Fe in the ground as well as in the excited state is given by [11]

$$\mathcal{H}_n = \langle \vec{S} \rangle \cdot \tilde{A} \cdot \vec{I} + \frac{eQV_{\hat{z}\hat{z}}}{4I(2I - 1)}$$

$$[3I_{\hat{z}}^2 - I(I + 1) + \eta(I_{\hat{x}}^2 - I_{\hat{y}}^2)] - \beta_n g_n \vec{B}^{ext} \cdot \vec{I} + \delta, \tag{21}$$

where the first term describes the magnetic hyperfine interaction, the second one, being present only in the nuclear excited state, the electric quadrupole interaction, the third one the nuclear Zeeman interaction, and δ represents the isomer shift. I is the nuclear spin taking on the values 1/2 and 3/2 for the ground and excited nuclear state of ^{57}Fe, respectively; eQ being zero in the ground state denotes the nuclear quadrupole moment of the excited state, $V_{\hat{z}\hat{z}}$ the main component of the electric field gradient (efg) tensor in the principal-axis system, which may be rotated against the reference frame of the electronic spin Hamiltonian, and η the asymmetry parameter ($\eta = (V_{\hat{x}\hat{x}} - V_{\hat{y}\hat{y}})/V_{\hat{z}\hat{z}}$ with $|V_{\hat{x}\hat{x}}| \leq |V_{\hat{y}\hat{y}}| \leq |V_{\hat{z}\hat{z}}|$); $\vec{B}^{ext}$ means the applied (external) field and β_n is the nuclear magneton. The nuclear g factor g_n and magnetic hyperfine coupling tensor $\tilde{A}$ depend both on the nuclear state, ground and excited, respectively, but the ratio $\tilde{A}/\beta_n g_n$, having the dimension of a magnetic induction, is to be regarded as constant. The spin operator in the first term has been replaced by its expectation value, because it is assumed that the electronic states are singlets with energy separations much larger than the hyperfine splittings. This can always be achieved by application of an external field of 2 mT or more. For the calculation of the resulting Mössbauer spectrum one has to keep in mind that there may occur also jumps between the various eigenstates of the electronic spin Hamiltonian. Either one of two possible approximations are usually made, depending on the condition whether these jump rates are slow or fast compared with the nuclear Larmor frequencies. In the first case, slow relaxation, each occupied spin state is assumed to contribute individually to the overall magnetic hyperfine pattern with a Mössbauer spectrum weighted by the corresponding Boltzmann factor. In the second case, fast relaxation, the thermal average of the spin expectation values is formed

$$\langle S_i \rangle_{av} = \sum_j \langle S_i \rangle_j \exp(-\varepsilon_j/kT)/\sum_j \exp(-\varepsilon_j/kT), \qquad (22)$$

where $i = x, y, z$, and j has to be summed over the $2S + 1$ substates of the ground-state spin multiplet. Entering these average values into formula (21) gives rise to one magnetic hyperfine field which produces one averaged magnetically split Mössbauer spectrum. In case that the jump rates between the electronic states are comparable with the nuclear Larmor frequencies, intermediate relaxation, a more sophisticated approach has to be used [12–15], which will be discussed in the context of the corresponding systems.

Polycrystalline samples or frozen solutions require furthermore an angular averaging over all directions in space, which is usually done numerically.

3.1.2 Spectrometers

The most usual implementation of iron Mössbauer spectroscopy is in transmission geometry using a single-line source and a split absorber [16]. In a typical transmission experiment the gamma rays emitted by the source are registered by a suitable detector after they have passed an absorber where they are partially absorbed. In order to investigate the energy spectrum of the nuclei it is necessary to modify the energy of the emitted gamma rays what is usually accomplished by moving the source relative to a stationary absorber. The resulting first-order Doppler shift is used to match the energy differences between the emitted line of the source and the absorption lines of the sample. The motion of the source is preferably oscillatory and synchronized with a multiscaler unit so that a chosen energy interval is repeatedly scanned many times.

A schematic diagram of a typical Mössbauer spectrometer is depicted in Fig. 13. The source motion is produced by an electromechanical transducer, similar to a loudspeaker, which is driven by a waveform generator. The detector system can be verified in the case of ^{57}Fe easily by an Ar filled proportional counter with standard nucleonics consisting of preamplifier, spectroscopic amplifier and single-channel analyser for selecting the Mössbauer gamma rays. The synchronization of the gamma ray counting and the source motion is generally achieved by a microprocessor system in which the counts are accumulated in channels corresponding to the velocity of the source relative to the absorber. The spectrum is collected for a period typically of the order of hours or days during which the spectrum may be monitored on a display screen. Calibration is performed with α-iron foils which give six-line spectra with well-known line positions.

The Doppler shift ΔE at a velocity v is given by

$$\Delta E = (v/c)E_0 \tag{23}$$

where E_0 is the energy of the Mössbauer transition. For the 14.4125 keV transition of ^{57}Fe one gets the following correspondencies

$$1 mms^{-1} \cong 4.8075 \times 10^{-8} \, eV \cong 3.8776 \times 10^{-4} \, cm^{-1} \cong 11.625 \, MHz.$$

The absolute isomer shifts observed depend on the matrix material of the source and are frequently quoted in older references with respect to the source

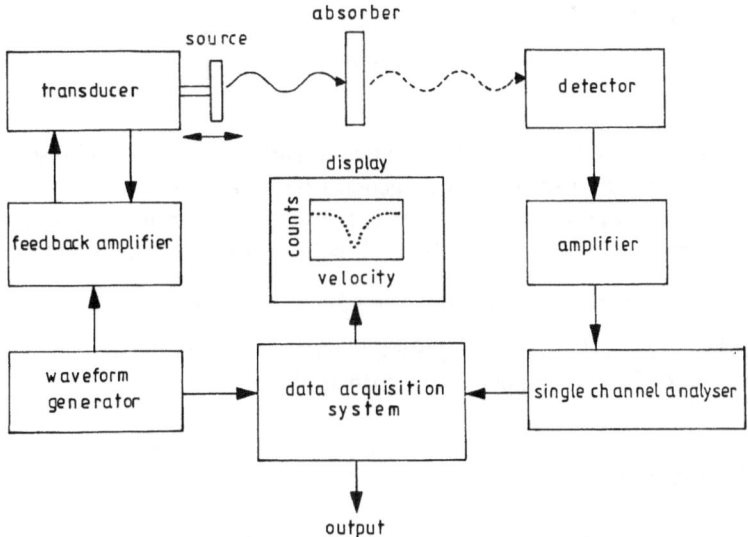

Fig. 13. Block diagram of a typical Mössbauer spectrometer

used. It is now customary to give the isomer shifts with respect to the inversion point of the α-iron spectrum taken at room temperature being thus independent of the source matrix.

When δ*(X) denotes the isomer shift of the sample with respect to a source in matrix X the isomer shift δ(α-Fe) with respect to α-Fe at RT is given by

$$\delta(\alpha\text{-Fe}) = \delta^*(X) + \delta_s(X) \tag{24}$$

where $\delta_s(X) = +0.106(2)\,\text{mms}^{-1}$ for Rh, $+0.177(2)\,\text{mms}^{-1}$ for Pd, $+0.225(2)\,\text{mms}^{-1}$ for Cu, and $+0.349(6)\,\text{mms}^{-1}$ for Pt [17].

3.2 EPR Spectroscopy

3.2.1 Principles

Electron paramagnetic resonance (EPR) spectroscopy observes transitions between magnetic dipole-moment eigenstates of a paramagnetic ion which are induced by high-frequency radiation. Since the energy separation between the states involved can be modified by the strength of the applied magnetic field, the optimum operating frequency with respect to sample characteristics as well as technology can be chosen. A good compromise between optimum sensitivity, optimum sample size, homogeneity, strength of the required field and component cost is achieved by operating in the microwave region of the electromagnetic spectrum. A field of about 0.35 T is needed to obtain resonance absorption at 10 GHz (X-band).

The position of the EPR signal which reflects the splitting of a pair of magnetic sublevels under the influence of an external magnetic field is characterized by the spectroscopic g value defined as

$$g = h\nu/\beta B_r \tag{25}$$

where ν is the spectrometer frequency and B_r is the resonance field. If monocrystalline, the sample will in general be anisotropic making g^2 – in the strict mathematical sense not necessarily g itself – a tensor quantity [7], which implies that the g value for an arbitrary direction of the magnetic field reads

$$g(\alpha, \beta, \gamma) = \sqrt{g_x^2 \cos^2 \alpha + g_y^2 \cos^2 \beta + g_z^2 \cos^2 \gamma} \tag{26}$$

with g_x, g_y, g_z denoting the principal components of the 'g tensor' and α, β, γ the angles of $\vec{B}$ with respect to the principal axes.

3.2.2 Spectrometers

An EPR spectrometer is always configured from an electromagnetic radiation excitation and detection system, a probe containing the sample, signal enhancement and signal processing electronics and a magnet. A functional block diagram of the basic ER Series Spectrometer from Bruker Analytics is provided in Fig. 14.

The microwave excitation and detection system, in conjunction with associated control electronics for the microwave components, is referred to as a microwave bridge. The microwave radiation is derived from a coherent radiation source such as a klystron. Frequency stability of the microwave source is achieved by locking the klystron frequency to the sample cavity frequency through a feedback loop. The sample cavity serves to contain the sample in the magnet air gap and dictates the operating frequency of the spectrometer. Its design and fabrication is paramount in determining the sensitivity, may limit stability and resolution and, in some cases, may determine whether a particular experiment can be performed or not. Cavities are characterized by their quality or Q factor which is essentially the ratio of the stored and dissipated energy per cycle or equivalently the ratio of resonant frequency and width of the resonance defined by the half-power absorbing points. The magnetic field dependence of the level splittings makes it possible to maintain the highest achievable Q factor and a constant cavity and source frequency. The spectrum is obtained by monitoring the absorption or dispersion EPR response as a function of the

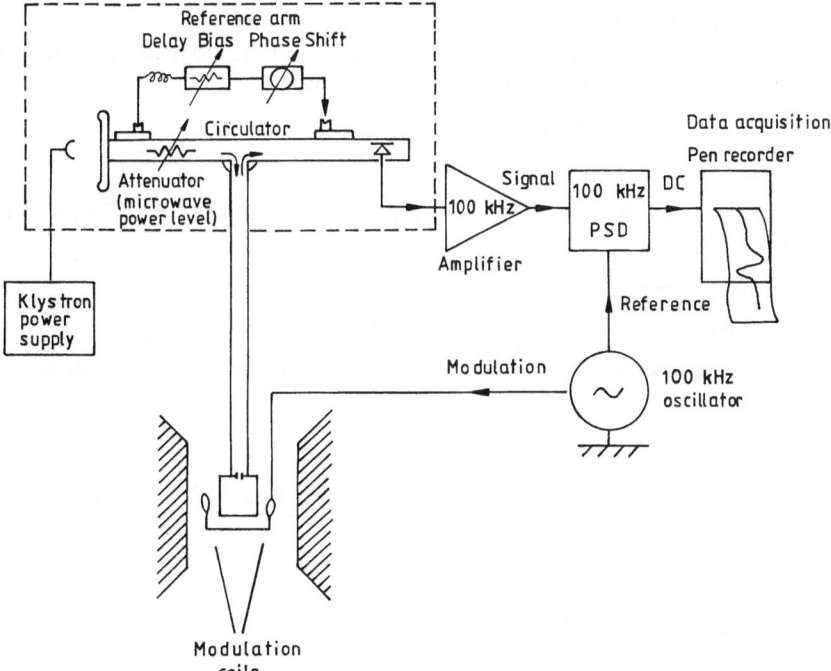

Fig. 14. Basic layout of an EPR spectrometer

applied magnetic field. The desired central field and the field sweep are monitored by the field controller with the help of a Hall device. The microwave detection system used in the ER Series EPR spectrometers works with a detector diode which is biased to the so-called square-law regime where the diode current is a linear function of power.

To provide amplitude modulation of the magnetic field all cavities are equipped with a set of Helmholtz coils which are attached to the signal channels. The field modulation detection scheme applied in the signal enhancement and processing channel improves the signal-to-noise ratio by amplifying only the response at the modulation frequency. The EPR signal response is decoded by phase sensitive detection, i.e. coherent with the magnetic field modulation, and as a consequence the output signals appear as derivatives of EPR absorption signals provided that (a) the modulation field is a small fraction of the EPR linewidth expressed in field units, (b) the frequency is a small fraction of the EPR linewidth expressed in frequency units, (c) the d.c. magnetic field is varied slowly through the resonance, and (d) no microwave saturation exists.

3.2.3 The Effective g Values for S = 5/2

Electron paramagnetic resonance requires the presence of unpaired electrons. This condition is naturally met by ions with an odd number of electrons. If such an ion is in an environment of symmetry sufficiently low for the ground state level to be isolated and have only twofold Kramers degeneracy, the Zeeman interaction of this doublet can be described by

$$\mathscr{H}_z^{\text{eff}} = \beta \vec{B} \cdot \tilde{g}^{\text{eff}} \cdot \vec{S}^{\text{eff}} \tag{27}$$

with $S^{\text{eff}} = 1/2$.

The ground state of the high-spin ferric iron has in the $S = 5/2$ spin space the form

$$|\phi_j^{\pm}\rangle = a_j^{\pm} |\pm 1/2\rangle + b_j^{\pm} |\mp 3/2\rangle + c_j^{\pm} |\pm 5/2\rangle \tag{28}$$

with coefficients $a_j^{\pm}, b_j^{\pm}, c_j^{\pm}, j = 1, 2, 3$, as given by formulas 5a–c. Establishing a one-to-one correspondence between the matrix elements of the Zeeman interaction term

$$\mathscr{H}_z = 2\beta \vec{B} \cdot \vec{S} \tag{29}$$

in the $S = 5/2$ representation and of $\mathscr{H}_z^{\text{eff}}$ in the three $S^{\text{eff}} = 1/2$ representations given by (5a–c), yields the relations

$$g_x^{\text{eff}(j)} = 4|\langle \phi_j^- | S_x | \phi_j^+ \rangle| \tag{30a}$$

$$g_y^{\text{eff}(j)} = 4|\langle \phi_j^- | S_y | \phi_j^+ \rangle| \tag{30b}$$

$$g_z^{\text{eff}(j)} = 4|\langle \phi_j^+ | S_z | \phi_j^+ \rangle| \tag{30c}$$

The effective g values depend only on the ratio E/D as it is shown in Fig. 15. This

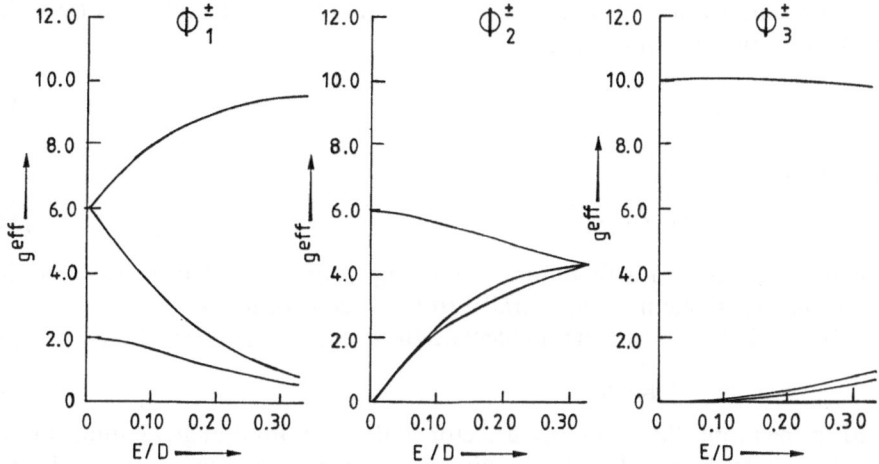

Fig. 15. Effective g values of the three Kramers doublets arising from the S = 5/2 multiplet plotted as a function of the rhombicity parameter E/D

demonstrates the ability of EPR spectroscopy to determine the sign of D and the rhombicity E/D from the three spectroscopic g values belonging to one EPR transition, for these are equivalent to the three effective g values of a Kramers doublet.

3.3 Magnetic Susceptibility

3.3.1 Principles

The magnetic susceptibility of a substance describes its response to magnetic fields. It is defined as the ratio of the magnetization M_H in the direction of the magnetizing field $\vec{H}$ and the strength of this field. For the molar susceptibility given in SI units it follows therefore [18]

$$\chi_H = (\mathcal{M}/\rho)\,M_H/H, \tag{31}$$

where $\mathcal{M}$ denotes the molar mass and ρ the mass density of the substance. The susceptibility is in general anisotropic and forms a real symmetric tensor, for which a principal-axis system (PAS) exists. For a field directed in an arbitrary direction χ_H can be calculated from its principal components $\chi_{xx}, \chi_{yy}, \chi_{zz}$ by means of the relation

$$\chi_H(\alpha, \beta, \gamma) = \chi_{xx} \cos^2 \alpha + \chi_{yy} \cos^2 \beta + \chi_{zz} \cos^2 \gamma, \tag{32}$$

where α, β, γ are the angles of $\vec{H}$ with respect to the PAS.

The magnetization being a macroscopic quantity can be obtained from the molecular magnetic moments μ through

$$M_H(T) = \frac{N_A \rho}{\mathcal{M}} \langle \mu_H \rangle_T,$$ (33)

and, thus,

$$\chi_H(T) = N_A \langle \mu_H \rangle_T / H,$$ (34)

where N_A is Avogadro's number and $\langle \mu_H \rangle_T$ the ensemble average of the molecular magnetic moment parallel to $\vec{H}$ at temperature T.

To continue one can either calculate the expectation values of the operator

$$\vec{\mu} = -(\vec{L} + g_s \vec{S})\beta,$$ (35)

with g_s denoting the spin-only g factor and $\vec{L}$ the total angular momentum, which requires knowledge of the molecular wave functions, or one can make use of the relation [19]

$$\mu_H = -\frac{\partial \mathcal{H}}{\partial B}$$ (36)

where only the derivative of the appropriate Hamiltonian with respect to B is needed. By denoting its eigenvalues by ε_j one obtains for the magnetic moment of the eigenstate $|\phi_j\rangle$

$$\langle \mu_H \rangle_j = -\frac{\partial \varepsilon_j}{\partial B}$$ (37)

and $\quad \langle \mu_H \rangle_T = -\dfrac{\sum\limits_j \dfrac{\partial \varepsilon_j}{\partial B} \exp(-\varepsilon_j/kT)}{\sum\limits_j \exp(-\varepsilon_j/kT)}.$ (38)

These moments are expected, of course, to depend on the direction as well as on the strength of the magnetizing field $\vec{H}$. If one assumes that ε_j can be expanded as power series in $B = \mu_0 H$

$$\varepsilon_j = \varepsilon_j^{(0)} + \varepsilon_j^{(1)} B + \varepsilon_j^{(2)} B^2 + \cdots$$ (39)

and consequently

$$\exp(-\varepsilon_j/kT) = \exp(-\varepsilon_j^{(0)}/kT)(1 - B\varepsilon_j^{(1)}/kT + \cdots),$$ (40)

one gets from formula (38) for the term linear in B, which is relevant in the absence of a permanent magnetization and far from saturation, the expression

$$\langle \mu_H \rangle_T = B \frac{\sum\limits_j [(\varepsilon_j^{(1)})^2/kT - 2\varepsilon_j^{(2)}] \exp(-\varepsilon_j^{(0)}/kT)}{\sum\limits_j \exp(-\varepsilon_j^{(0)}/kT)}$$ (41)

This is the well-known Van Vleck formula [19].

The spectroscopic g values determined by EPR spectroscopy can be easily related to the molecular magnetic moments if $h\nu$ in formula (25) is identified with $2|\varepsilon_j - \varepsilon_j^{(0)}|$. As long as ε_j depends only linearly on B (37) yields

$$|\langle \mu_H \rangle| = \tfrac{1}{2} g\beta. \tag{42}$$

This means for the three Kramers doublets of ferric high-spin iron $|\phi_j^{\pm}\rangle$, $j = 1, 2, 3$, which are given explicitly by expression (5a–c), that

$$|\langle \mu_z \rangle_j| = \tfrac{1}{2} g_z^{\mathrm{eff}(j)} \beta, \tag{43a}$$

when B is directed along the z direction. When $\vec{B}$ is, on the other hand, directed perpendicular to the z direction appropriate linear combinations of $|\phi_j^+\rangle$ and $|\phi_j^-\rangle$ are basis states yielding

$$|\langle \mu_{x,y} \rangle_j| = \tfrac{1}{2} g_{x,y}^{(j)} \beta. \tag{43b}$$

The $g_{x,y,z}^{(j)}$ are those plotted in Fig. 15 as a function of E/D.

If the magnetization is purely of orientational origin the Langevin formula for the molar susceptibility

$$\chi^M(T) = \mu_0 N_A \frac{n_{\mathrm{eff}}^2 \beta^2}{3kT} \tag{44}$$

is obtained in the high-temperature limit, where n_{eff} is the so-called 'effective Bohr magneton number' [19]. It is therefore customary to quote results of magnetic susceptibility measurements generally in terms of the effective magnetic moment $\mu_{\mathrm{eff}} \equiv n_{\mathrm{eff}} \beta$, which is according to (44) related to the molar susceptibility χ^M by

$$\mu_{\mathrm{eff}} = \sqrt{3kT\chi^M/(\mu_0 N_A)}. \tag{45}$$

For a state with spin S the spin-only value is expected to be

$$\mu_{\mathrm{eff}} = 2\beta\sqrt{S(S+1)}. \tag{46}$$

3.3.2 Susceptometers

The determination of the magnetic susceptibility is based, in general, on effects produced by the macroscopic moment of the sample [20]. These can be either the force exerted by a magnetic field gradient and measured, e.g. in a Faraday balance or the electric signal induced in a coil when the sample is moved. A typical device which uses the second principle is the Foner susceptometer. The most modern systems work with a SQUID (superconducting quantum interference device). Since this method is likely to become standard in future, some details of practical importance shall be presented here.

In Fig. 16a it is shown how the pick-up coils and the superconducting solenoid are arranged in the MPMS (Magnetic Property Measurement System) of Quantum Design. The coils are wound in a second-derivative configuration

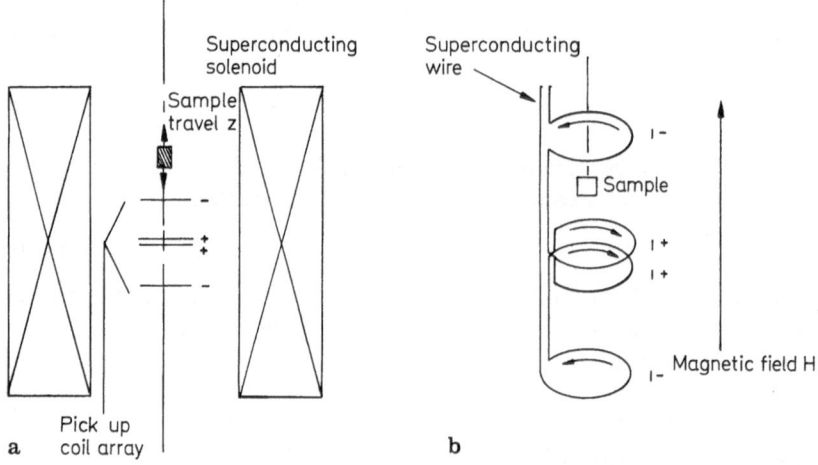

Fig. 16a, b. The magnet configuration (**a**) and the sensor (second-derivative coil) configuration (**b**) of the MPMS SQUID-susceptometer

(Fig. 16b) in which upper and lower single turns are counterwound with respect to the two-turn center coil. This configuration strongly rejects interference from nearby magnetic sources, and allows the system to function without any benefit from a superconducting shield. The normal measurement technique used in the MPMS system is to position the sample below the sensing coils far enough so that the sample moment is not detected and then to raise the sample through the coils. The signals induced in the second-derivative coils are coupled into a SQUID sensor through an isolation transformer. The sample is then measured by repeatedly moving it upward some distance and reading the voltage from the SQUID detector. If the SQUID voltage has been read at a large number of points, the values can be stored as a function of sample position. At each position the voltage is typically read several times, and many vertical scans can be averaged to improve the measurement resolution. Finally, the magnetic moment of the sample is computed numerically from the size of the signal in comparison with a NBS (National Bureau of Standards) platinum calibration sample. With the pick-up coil geometry used in the MPMS, samples of the size of the NBS standard, which forms a circular cylinder approximately 2.5 mm in diameter by 2.5 mm high, can be regarded effectively point sources to an accuracy of about 0.5 percent. For larger sample sizes a correction has to be applied to the data.

3.4 Comparison of the Three Methods

In the preceding sections it has become evident that the results obtained by the three techniques supplement and support each other. The combined information enables the spectroscopist to contrive a consistent picture of the magnetism

Table 1. Experimental conditions for the three techniques

	Mössbauer (^{57}Fe)	EPR (X-band)	SQUID
Maximum volume of sample	1 cm^3	0.1 cm^3	0.1 cm^3
Minimum amount of paramagnetic species	21 μ mole[a]	1 μ mole	100 μ mole[b]
Typical time for measurement	0.5–5 d	0.1–1 h	1 min (8 h for temp scan 2–300 K)

[a] With natural isotope abundance of iron [b] in solution

of the corresponding transition metal complex. In practice it turns out that the experimental requirements, in particular for the sample concentration, are rather different. In Table 1 a survey is given, which is meant to represent a standard situation and not the absolute limit attainable by extreme efforts.

4 Mononuclear Iron Complexes

In Sects. 4 and 5 we will present examples of spin-Hamiltonian analyses of EPR, Mössbauer and susceptibility data obtained from iron proteins and related synthetic analogs. The examples are described in the order of increasing complexity of the interactions required in the spin Hamiltonian parametrization. In Sect. 4 we are dealing with mononuclear iron complexes. We briefly mention complexes without zero-field splitting, treat systems with zero-field splitting in axial as well as rhombic ligand-field symmetry and finally discuss examples of sizeable spin admixture, which require extensions of the usual spin Hamiltonian formalism. In Sect. 5 we enter the field of homo and hetero di- and oligo-nuclear paramagnetic clusters which exhibit spin coupling.

4.1 Iron Complexes Without Zero-Field Splitting

The magnetic properties of complexes with an energetically well separated ground spin-doublet can be described by the spin Hamiltonian (Eq. 1) with $S = 1/2$, which in this case reduces to $\mathscr{H}_e = \beta \vec{S} \cdot \tilde{g} \cdot \vec{B}$. According to Kramers theorem the doublet state is not split by zero-field interaction, therefore only the Zeeman splitting remains in the spin Hamiltonian.

4.1.1 Low-Spin Ferric Iron ($S = 1/2$)

The low-spin ferric porphyrins are the only examples of $S = 1/2$ iron-complexes of biological relevance. In fact, with a few exceptions [21] only hemes, cyto-chromes and synthetic porphyrins with hexa-coordination of the ferric iron and with strong axial ligands provide a strong ligand field and therefore low-spin conditions.

The g-tensor in low-spin ferric iron is usually interpreted in the framework of a ligand-field model [22, 23]. The ligand-field parameters, deduced from experimental EPR g-values, were exploited in correlation diagrams for classification of axial iron ligands [24]. A detailed review on EPR analyses of low-spin porphyrins is given in ref. [25].

The ligand-field model was also applied to the interpretation of Mössbauer spectra of low spin ferric hemes [26, 27]. However, systematic Mössbauer simulations demanded extensions of the initial model, for instance to allow for rotations between the principal axes of the g-, A- and the efg tensors [28]. A detailed review of Mössbauer studies on low-spin ferric porphyrins is given in Ref. [5].

Magnetic susceptibility studies of low-spin ferric systems are of little significance, because the anisotropy of $\tilde{g}$ can hardly be unravelled from macroscopic magnetic moments [29]. Mainly room-temperature values are reported for diagnostic purposes.

4.2 Zero-Field Interaction with Axial Symmetry

4.2.1 Iron (S = 5/2) in Porphyrins

High-spin ferric iron porphyrins are typical examples of iron complexes with axial ligand field symmetry. The iron is penta-coordinated with significant displacement out of the porphyrin plane towards the axial ligand. Hexacoordination may also offer high-spin conditions in ferri-porphyrins provided the axial ligands are weak [30]. EPR, Mössbauer and magnetic susceptibility data of the Fe^{3+} ion are usually analyzed on the basis of a spin Hamiltonian for S = 5/2 (Eq. 1).

4.2.1.1 EPR Spectroscopy

The presence of zero-field splitting in ferric iron porphyrins can be visualized directly from the effective g-values. Figure 17 shows a typical low-temperature EPR spectrum of a synthetic iron porphyrin, with resonances at $g^{eff} = 6$ and $g^{eff} = 2$. The effective g-values can be explained by the properties of a low lying, thermally occupied Kramers doublet Φ_1, as presented in Fig. 1 and Eq. 5a. The spin expectation values of this doublet, generated by an applied field B, depend strongly on the orientation of the field in the molecular axes system, as shown in Fig. 2. For vanishing rhombicity $E/D \sim 0$ the spin expectation values of the lowest doublet Φ_1 are $|\langle S_{x,y} \rangle| = 3/2$, $|\langle S_z \rangle| = 1/2$. The corresponding effective g-values are derived from Eq. (30a–c) as 6., 6. and 2. (Fig. 15), in accordance with experiment.

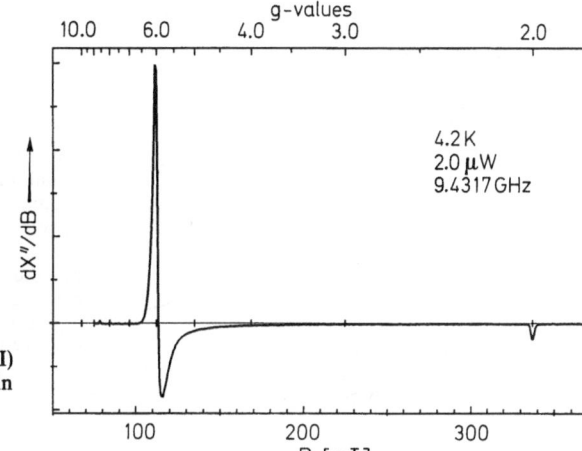

Fig. 17. EPR spectrum of iron(III) picket-fence porphyrin dissolved in toluene

We note that excited doublets do not contribute to the EPR spectrum in the case of strong axial zero-field interaction ($|D| > g\beta B$, $E = 0$). Under this condition, Eq. (7a–c), mixing of wave functions is negligible and therefore transition probabilities vanish ($\Delta M > 1$). Furthermore, inter-doublet transitions do not occur, provided the corresponding energy differences are larger than the microwave energy $h\nu$, which is usually the case in iron(III) porphyrins ($D > 3.5$ cm^{-1}, $h\nu \sim 0.3$ cm^{-1} in X-band EPR).

If, on the other hand, in our example the zero-field splitting were negligible with respect to the Zeeman interaction, then the magnetic field of the EPR spectrometer would split the $|S = 5/2, M\rangle$ spin sextet into six equally spaced levels, for any orientation of the field (see Sect. 2.1.2). As a consequence, all EPR resonances would be expected at $g^{eff} = g_e$ ($= 2.003$), which is not observed for high-spin ferric iron in porphyrins.

Figure 15 shows, for $S = 5/2$ systems, how effective g-values correlate with rhombic distortions. If the rhombicity is only a small perturbation, E/D-values can be easily determined from the first order perturbation expression for Φ_1 ($S = 5/2$): $g^{eff}_{x,y} = 6 \pm 24 \cdot E/D$, $g^{eff}_z = 2$, using apparent splittings of the $g^{eff}_\perp$ resonances. High-spin ferri-porphyrins usually show weak or even negligible splitting of the $g^{eff}_\perp$ resonances, as in Fig. 17, and hence, E/D is close to zero.

The thermal population of Kramers doublets, detectable from the temperature dependence of the EPR intensities [31], allows one to probe the zero-field parameter D. The EPR intensity is proportional to $p_1^- - p_1^+$, the difference in thermal weights of the sublevels Φ_1^+ and Φ_1^- of the resonating doublet. Expansion of $p_1^- - p_1^+$ yields in first order a Curie-Weis law, modified by the Boltzmann population of the doublet. Hence, plotting the 'intensity I times temperature T' as a function of temperature reveals the corresponding Boltzmann function, which depends on D. In temperature ranges, where the line widths are sufficiently constant, the integrated intensities can be replaced simply

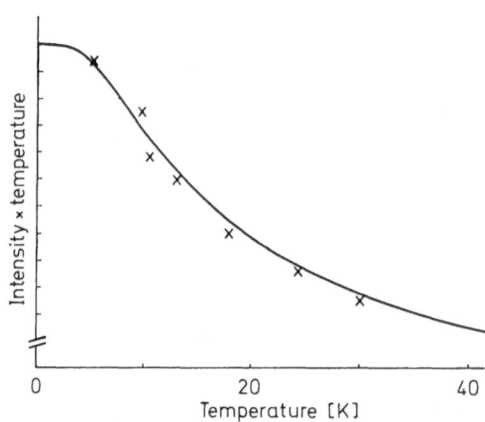

Fig. 18. Temperature dependence the $g_\perp$ signal of prostaglandin-H-synthase. The ordinate represents "intensity of the derivative signal times temperature". The *solid line* is a fit of the thermal depopulation of the ground doublet $p(T) = (1 + \exp(-2D/kT) + \exp(-6D/kT))^{-1}$, yielding a value $D = 6.1 \pm 1.5\,cm^{-1}$. Taken from [40]

by the intensities of the derivative signals. This procedure is comparable to measurements of the static magnetic susceptibility and has approximately the same accuracy, however, with the advantage that contributions from different paramagnetic centers can be discerned. An application is illustrated in Fig. 18 for the ferric iron in the heme enzyme prostaglandin-H-synthase (PGH-synthase) [32, 33, 34].

Moderate values of D ($\sim g\beta B$) can also be determined by EPR from the competition of zero-field interaction with Zeeman interaction (Sect. 2.1.3). The magnetic perturbation of the spin multiplet by the resonance field causes mixing of wave functions, depending on D and $g\beta B$. In turn, in an EPR experiment the resonance field depends on the microwave frequency. As a result, the effective g-values become a function of microwave frequency and zero-field splitting [35]. For porphyrins, however, D is usually too large to allow for sizeable effects. However, we will show that in Mössbauer spectroscopy, where strong fields up to several tesla can be applied, also these larger D-values can be determined from the effects of magnetic perturbations.

Zero-field splittings can also be determined from temperature-dependent EPR measurements of spin-lattice relaxation time T_1 [36, 37]. Under the condition that Orbach processes govern the longitudinal relaxation, T_1 varies with the thermal population of excited Kramers doublets [38]. However, the prerequisites for unique dependence of the EPR amplitudes on the spin-lattice relaxation time, like perfect magnetic dilution of the samples with well isolated paramagnetic centers have to be met.

Of particular importance in any temperature-dependent evaluation of zero-field splittings are accurate temperature measurements and control. In the EPR measurements on PGH-synthase the Curie-Weiss behavior of a ferric low-spin signal (S = 1/2) was used to derive an internal standard for temperature calibration [39] in a usual helium-flow cryostat with standard cavity [40]. To this end, a sample of PGH-synthase was "overlayed" in the EPR tube with a few microliters of ferric low-spin cytochrome c, in order to get both types of resonances simultaneously in the spectrum, those from PGH-synthase and, well resolved, also the g = 3 signal of low-spin cytochrome c. From a plot of the intensity of the low-spin signal the temperature scale was calibrated in the above analysis of the thermal depopulation of the lowest Kramers doublet of iron(III) in PGH-synthase.

4.2.1.2 Mössbauer Spectroscopy

The characteristic features of low-temperature Mössbauer spectra of S = 5/2 complexes with quasi axial zero-field interaction are illustrated by the spectra of ferric PGH-synthase (Fig. 19). At 4.2 K in applied fields the spectra exhibit magnetic hyperfine interactions due to slow spin relaxation. The overall splitting of about 9.5 mms^{-1} (at 20 mT applied field) corresponds to a hyperfine field sensed by the iron nucleus of about 30 T.

Within axial ligand field the anisotropy of spin expectation values of the low lying Kramers doublet ϕ_1, as it was probed by the g-values in the EPR experiments of PGH-synthase, describes an easy plane of magnetization, spanned by the x- and y-axes. This means that for arbitrary oriented external fields the main component of the spin expectation value $\langle \vec{S} \rangle$ and the hyperfine field $\vec{B}^{hf} = -\tilde{A} \cdot \langle \vec{S} \rangle / g_n \beta_n$ at the ^{57}Fe nucleus tend to be aligned along the projection of the field within the xy-plane. Substituting the spin expectation values $|\langle S_{x,y} \rangle| = 3/2$ and the experimental value $B^{hf} \sim 30$ T in the above expression, yields $A_{x,y}/g_n \beta_n \sim 20$ T, which is in the usual range for high-spin ferric iron.

Due to the anisotropy of the spin expectation values, the components A_x and A_y have more influence on the simulations and therefore are determined more accurate than A_z. Only in strong applied fields the spin tends to follow the field orientation, and then A_z gains more influence. Spin Hamiltonian simulations of the experimental lineshapes at a field of 6 T yielded for PGH-synthase a value for A_z which is larger in magnitude than that of $A_{x,y}$. The analysis could not establish a difference in A_x and A_y.

The magnetic anisotropy of the ground Kramers-doublet has the consequence, that for a random arrangement of molecules in a powder sample (frozen solution) the internal fields are not randomly oriented in space but show strong texture around the axis of the external field. In fact, the relative line intensities of the Mössbauer line-sextet are found at ratios close to 3:4:1:1:4:3 for weak fields applied perpendicular to the γ-rays, instead of 3:2:1:1:2:3 expected for random orientation of hyperfine fields [41]. In this estimate of relative Mössbauer intensities we make use of the fact, that the electric quadrupole interaction in this case is only a perturbation of the magnetic hyperfine interaction. For completeness we note that in cases where the electronic ground doublet exhibits an easy axis of magnetization the alignment of spin expectation values in a powder sample is random with respect to the γ-beam. For instance, the $S_z = \pm 5/2$ Kramers doublet Φ_3 (Fig. 15) has such an easy axis of magnetization ($g^{eff} = (10, 0, 0)$ for E/D = 0). If Φ_3 is ground state, the intensity ratios of

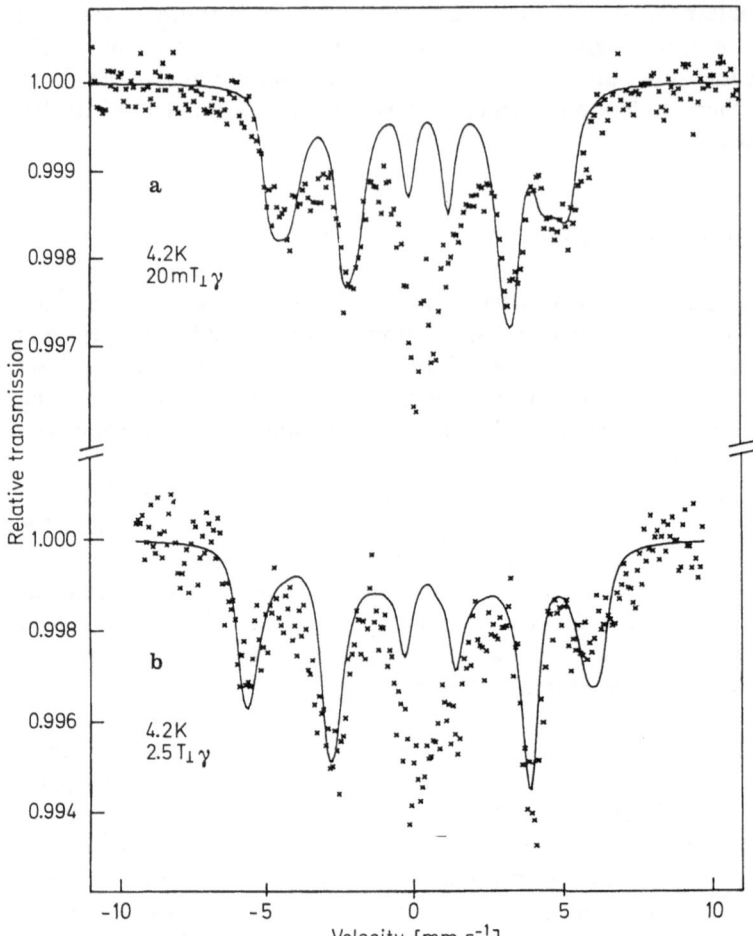

Fig. 19. Mössbauer spectra of ferric prostaglandin-H-synthase at 4.2 K in fields of 20 mT (**a**) and 2.5 T (**b**), applied perpendicular to the γ-rays. The solid lines are spin-Hamiltonian simulations using $D = 5.6 \text{ cm}^{-1}$, $E/D = 0.026$, $\tilde{A}/g_n\beta_n = (-19.4, -19.4, -22.)$ T, $\Delta E_Q = +0.9 \text{ mms}^{-1}$, $\eta = 0.2$, $\delta(\alpha\text{-Fe}) = 0.36 \text{ mms}^{-1}$. The deviations between theory and experiment are due to non-specifically bound hemin in the protein. Taken from [40]

the six Mössbauer lines are $3:2:1:1:2:3$ independent on the orientation of weak fields with respect to the γ-beam.

As mentioned above, the zero-field splitting parameter D can be determined from the response of spin expectation values on perturbations by magnetic fields. Accordingly, the functional dependence of the hyperfine field on the strength of the applied field contains information about the zero-field splitting. From the Mössbauer spectra of PGH-synthase, recorded with external fields of 20 mT and 2.5 T, a value $D = 5.6 \pm 1 \text{ cm}^{-1}$ was found from spin-Hamiltonian simulations by adjusting D and A. The value is in good agreement with the EPR result ($D = 6.1 \pm 1.5\ cm^{-1}$).

The components of the efg tensor at the iron nucleus in PGH-synthase could not be determined directly from the quadrupole splitting of a high-temperature Mössbauer spectrum. The magnetic splitting did not collapse sufficiently and in addition, the superposition of specific and adventitious heme subspectra prohibited a simple analysis [40]. In low-temperature Mössbauer spectra, the electric quadrupole interaction appears as a perturbation of the nuclear Zeeman interaction. In first-order approximation the Mössbauer spectra depend only on the components of the efg tensor along the hyperfine field, i.e. V_{xx} and V_{yy} in the present case. Therefore, the apparent quadrupole splitting in the magnetic spectra (the shift of the inner four lines vs. the outer two) is opposite in sign and only half the value $\Delta E_Q = +0.9\ mms^{-1}$ given in the caption of Fig. 19.

A comprehensive review of Mössbauer and EPR studies of porphyrin systems are found in the Physical Bioinorganic Chemistry Series [5]. Separate reviews on EPR techniques and magnetic susceptibility measurements in this field can be found in the same series [25, 29].

4.2.2 Non-Heme Iron (S = 3/2)

Studies of various non-heme iron-enzymes show that the ligand geometry may change within their catalytic cycle. Substrate binding can significantly modulate the protein conformation including the coordination sphere of iron. The symmetry of specifically bound and catalytically active ferric iron species deviates in most cases significantly from the rhombic limit $E/D = 1/3$, in contrast to non-specifically (adventitiously) coordinated ferric iron in biomolecules. We focus on the example of the chemically stable NO complex of the non-heme cofactor iron of the monooxygenase putidamonooxin (PMO) [42, 43, 44]. We shall show that it can be regarded as a model for an $iron \cdot O_2$ complex, which is a transient catalytic intermediate of the enzyme reaction of PMO. Because of the fast kinetics of the enzymatic process, the 'iron $\cdot$ peroxo' complex, which is the active oxygenating species, could not be detected by spectroscopic techniques.

4.2.2.1 EPR Spectroscopy

The EPR spectra of nitrosylated PMO in presence of the physiological substrate 4-methoxybenzoate (Fig. 20) has at 3.6 K resonances at $g^{eff} = 4.1$, 3.96, 2.0 [44], which are characteristic of a $|M_s = \pm 1/2\rangle$ Kramers doublet of spin $S = 3/2$ in axial ligand field with weak rhombic distortion. From the first-order perturbation expression $g^{eff}_{x,y} = 4 \pm 6 \cdot E/D$ we estimate $E/D = 0.01$. We note, that signals of paramagnetic Fe-S centers in the protein are saturated under the measuring conditions of Fig. 20.

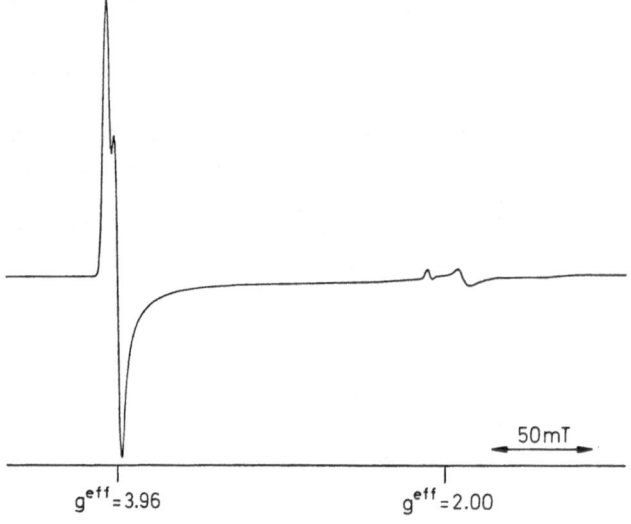

Fig. 20. EPR spectrum on nitrosylated putidamono-oxin in presence of 4-methoxybenzoate at 3.6 K. Taken from [44]

50 mT

$g^{eff} = 3.96$ $g^{eff} = 2.00$

4.2.2.2 Mössbauer Spectroscopy

PMO contains one [2Fe-2S] cluster and one mononuclear non-heme iron (cofactor) per active center. The protein can be reversibly depleted from its cofactor iron and reconstituted with ferrous iron under reducing conditions in the presence of substrate [43]. By combinations of in vivo enrichment with ^{57}Fe of the [2Fe-2S] cluster and in vitro enrichment of the cofactor site we were able to prepare specific Mössbauer samples for investigating the sites separately. For the study of the nitrosyl complex we are concerned with $[2^{56}Fe\text{-}2S]$ + cofactor ^{57}Fe. The spectra obtained at 4.2 K in moderate fields, applied parallel and perpendicular to the direction of the γ-radiation, are shown in Fig. 21. The relative intensities of the prominent six-line subspectrum are strongly dependent on the orientation of $\vec{B}^{ext}$. This finding is compatible with an easy plane of magnetization within the ground state. From this we conclude that D is positive and the $|3/2, \pm 1/2\rangle$ Kramers doublet is lowest in energy. The actual value of the zero-field splitting $D = 12 \, cm^{-1}$ was obtained from the saturation behavior of the magnetic splitting of spectra recorded in applied fields in the range 100 mT − 6.4 T.

The hyperfine coupling tensor $\tilde{A}$ could be taken as isotropic in the spin-Hamiltonian simulations. However, since the z-components of the spin expectation values are relatively small the component A_z has little effect on the Mössbauer spectra and remains inaccurate in the analysis, relative to the A_x and A_y values, like for the $M_s = \pm 1/2$-doublet of spin $S = 5/2$, mentioned above. Therefore, an anisotropic, "disc-shaped" A-tensor, as it was predicted from calculations for model complexes [45, 46], is compatible with experiments.

We note that also the electric quadrupole splitting $\Delta E_Q = -1.4 \, mms^{-1}$ and the isomer shift $\delta(\alpha\text{-Fe}) = 0.68 \, mms^{-1}$ were obtained from magnetically per-

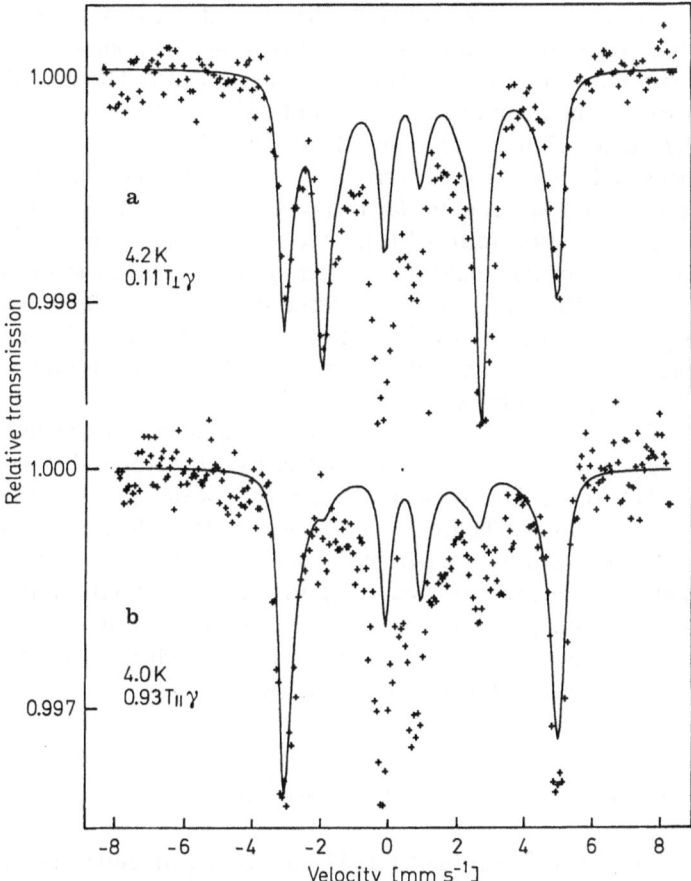

Fig. 21. Mössbauer spectra of the nitrosylated cofactor iron of putidamonooxin in the presence of 4-methoxybenzoate, (**a**) at 4.2 K and $\tilde{B}^{ext}(0.11$ T) applied perpendicular to the γ-rays and (**b**) at 4.0 K and $\tilde{B}^{ext}(0.93$ T) applied parallel to the γ-rays. The *solid lines* are spin-Hamiltonian simulations using $D = 12 \text{ cm}^{-1}$, $E/D = 0.01$, $\tilde{A}/g_n\beta_n = -23.5$ T (isotropic), $\Delta E_Q = -1.4 \text{ mms}^{-1}$, $\eta = 0.2$, $\delta(\alpha\text{-Fe}) = 0.68 \text{ mms}^{-1}$. The sample was contaminated with adventitiously bound iron. Taken from [43]

turbed spectra, because the analysis of the zero-field spectra was hampered by nonresolved overlapping subspectra from adventitiously bound iron.

4.2.2.3 Discussion of the S = 3/2 Ferric State

The absence of hyperfine splitting from ^{14}N in the EPR spectra of nitrosylated PMO reveals negligible spin density on NO [44]. This, in conjunction with the drastic decrease of the isomer shift, due to NO binding to ferrous iron, by more than 0.5 mms^{-1} [43] leads to the conclusion, that an exchange-coupled system of ferrous iron (S = 2) and NO radical (S = 1/2) can not explain the origin of the observed spin S = 3/2. However, two other configurations, $Fe(d^5)NO^-$ and $Fe(d^7)NO^+$, can also provide spin states S = 3/2. The d^5 and the d^7 configurations of iron are found in a variety of compounds which exhibit isomer shifts in the range 0.2–0.6 mms^{-1} for d^5 and 1.7–2.00 mms^{-1} for d^7 [41]. Since the measured value is 0.68 mms^{-1}, it was concluded that the iron in the NO complex of PMO is in the ferric S = 3/2 spin state. This conclusion is corrroborated by titration results [44], which show that the nitrosylated cofactor iron cannot be further oxidized by ferricyanide. Hence the formal valence in the model complex $Fe^{3+} \cdot NO^-$ corresponds to that of the native, highly unstable oxy complex $Fe^{3+} \cdot O_2^-$. This means that the NO complex allowed to study the ligand structure of the cofactor iron in a state, which is analogous to that of the active oxygen complex.

For the interpretation of the spectroscopic data of the NO complex of PMO results from other S = 3/2 compounds, e.g. from iron-dithiocarbamates X [45, 47, 48, 49], iron-dithiooxalates X [50], with X = halogen, and iron (5-Cl-salen) NO [51], were used. The ferric S = 3/2 spin state of the iron in these compounds is associated with square pyramidal penta-coordination. It is tempting to assume that a ligand field corresponding to penta-coordination also causes the ferric S = 3/2 state in the NO complex of PMO. This assumption includes that, upon NO binding to the reduced cofactor iron (which is penta- or hexa-coordinated [43]), at least one ligand has to be released.

4.3 Rhombic Zero-Field Interaction

The rhombicity parameter E/D reflects the inequivalence of the ligand-field along the molecular x- and y-axes, perpendicular to the tetragonal symmetry axis. The analysis of spectroscopic data from systems with rhombic distortions is illustrated by iron-sulfur proteins and related analogs.

4.3.1 Ferric Iron in FeS$_4$ Centers

In iron-sulfur proteins the Fe ions are located at the center of distorted sulfur tetrahedra [52]. From their chemical nature the iron ligands can be cysteinyl sulfur atoms and/or inorganic sulfide ions. In the Rieske center the cysteine residues are partially replaced by histidine [53, 54]. Depending on the protein, the active centers are monomeric units or linked clusters of 2, 3 or 4 tetrahedra. The iron-sulfur proteins are involved in many biological redox processes. The simplest representatives are the rubredoxins [52] and desulforedoxin [55], which contain a mononuclear center per protein monomer. The spectral properties of the iron-sulfur proteins are significantly affected by the rhombic distortions of the metal site.

4.3.1.1 EPR Spectroscopy

An EPR spectrum of ferric desulforedoxin, taken at 4.8 K is presented in Fig. 22. Fitting the g-values of the prominent resonances to the curves of Fig. 15 reveals significant deviation from axial symmetry ($E/D \sim 0.1$). The splitting of the x- and y-resonances reflects the rhombic distortion. The inset of Fig. 22 shows the three Kramers doublets in conjunction with theoretical effective g-values obtained with positive D and $E/D = 0.08$. It can be seen that the simple parametrization reproduces the observed resonances quite well: $g_y^{eff} = 7.7$, $g_x^{eff} = 4.1$ and $g_z^{eff} = 1.8$ for the ground doublet and $g_z^{eff} = 5.7$ for the first excited doublet.

We remark that the ferric FeS_4 units of the rubredoxins have a higher rhombic distortion than those of desulforedoxin, e.g. $E/D = 0.28$ for rubredoxin from *C. Pasteurianum* [56]. In these complexes the effective g-tensors are close to the values for maximal rhombicity $E/D = 1/3$: $g_1^{eff} = (0.86, 9.68, 0.61)$, $g_2^{eff} = (4.29, 4.29, 4.29)$ and $g_3^{eff} = (0.86, 0.61, 9.68)$ (compare with Fig. 15). In this limiting case the EPR spectra are dominated by strong signals at $g^{eff} = 4.3$, originating from the isotropic middle doublet, and a single peak near $g^{eff} = 9.7$ [31].

It is interesting to remember, that the rhombic distortions mix the M_s-sublevels of the spin multiplet (see eq. (5a–c)), which enables $|\Delta M| = 1$ EPR-transitions to take place in each of the Kramers doublets. The doublets contribute to the experimental EPR spectrum according to their thermal weights, which depend on zero-field splitting and temperature. In order to determine the zero-field splitting of desulforedoxin, the thermal occupation in the lower and the middle Kramers doublet were quantified from EPR spectra measured at 4.8 K [55]. The approximative formulas of Aasa and Vanngard [57] were used to reveal the total integrated intensities of the subspectra from

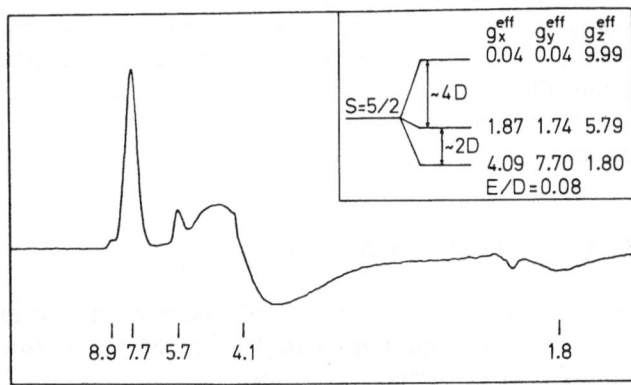

Fig. 22. EPR spectrum of ferric desulforedoxin at 4.8 K, microwave power 0.2 mW, frequency 9.24 GHz, modulation amplitude 1 mT. The inset gives the energy level scheme and theoretical g values for g = 2.0 and E/D = 0.08. Taken from [55]

integrations of isolated derivative peaks. This assessment yielded a value $D = 2$ cm^{-1}. In other cases of biological samples, where inhomogeneities of the iron moiety led to distorted line shapes, spin-Hamiltonian simulations including distributions of spin-Hamiltonian parameters were necessary to achieve reliable EPR quantifications and zero-field splittings [5, 58].

4.3.1.2 Mössbauer Spectroscopy

The EPR value of the zero-field parameter D in desulforedoxin was corroborated by the Mössbauer result ($D = 2.2 \pm 0.3$ cm^{-1}), derived from simulations of spectra recorded at 4.2 K in applied fields of 0.05 T, 1 T, 2 T and 4.5 T. The weak-field Mössbauer spectrum (50 mT, 4.2 K) exhibited rather broad features, in contrast to the strong field spectrum (4.5 T, 4.2 K). The broadness is attributed to the anisotropy of spin expectation values in the ground state, which are $\langle \vec{S} \rangle = (-1.92, -1.03, -0.45)$ according to the effective g-values. At 4.5 T, level mixing produces almost equal spin expectation values in x- and y-directions, leading to well resolved Mössbauer pattern [55].

In oxidized rubredoxins, the large rhombicity manifests itself at low temperatures by the Mössbauer features of an easy axis of magnetization. The internal fields tend to be aligned along the molecular y-axis for $D > 0$ and along the z-axis for $D < 0$, both being equivalent if $E/D = 1/3$. As a result, the orientation of the hyperfine fields is pinned relative to the principal axis system of the efg, yielding narrow lines in the Mössbauer spectrum.

4.3.2 Ferrous Iron in FeS$_4$ Centers

Rhombic zero-field interactions were also studied on a series of high-spin ferrous Fe–S$_4$ complexes [52, 59]. Synthetic analogs allowed to combine spectroscopic investigations, also on monocrystals, with X-ray structure determinations. In the high-spin ferrous compounds ($S = 2$) the large zero-field interaction usually inhibits EPR transitions. However, because of the large D-values, magnetic susceptibility measurements are an expressive tool for the study of Fe^{2+}S$_4$ clusters.

4.3.2.1 Mössbauer Spectroscopy

Iron in reduced rubredoxins, desulforedoxin and synthetic analogs exhibits large quadrupole splittings $|\Delta E_Q| > 3$ mms^{-1} and isomer shifts $\delta(\alpha$-Fe$) \sim 0.7$ mms^{-1}, characteristic of ferrous high-spin iron ($S = 2$) in tetrahedral sulfur ligation [5, 52]. In zero magnetic-field the spin expectation values are zero for each of the five spin states of the $S = 2$ multiplet, and the values increase linearly with B for small fields and finally reach a saturation value (Sect. 2.2.1). The

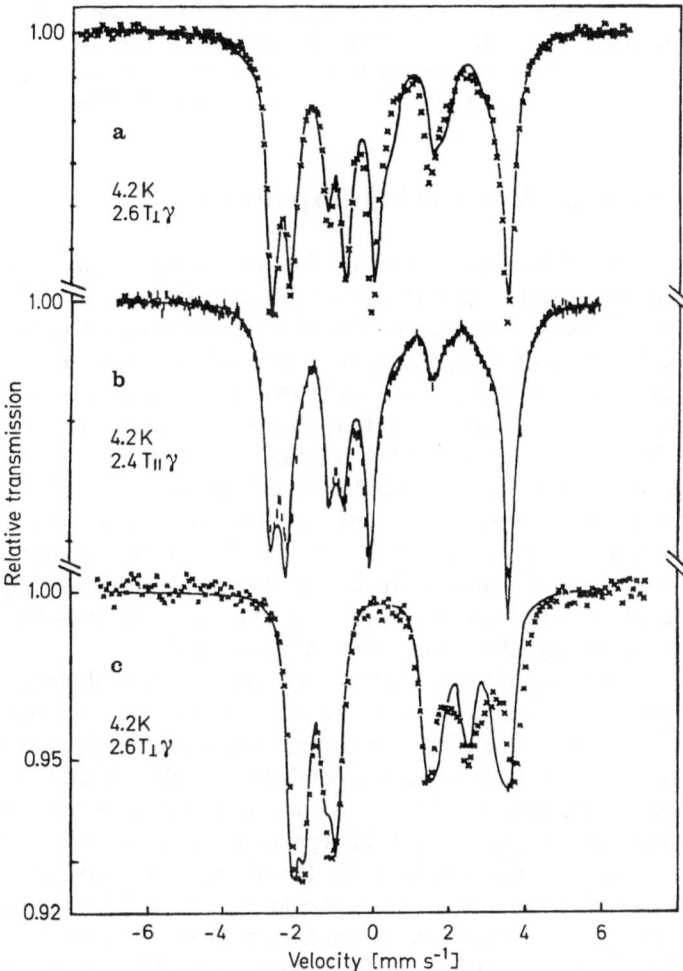

Fig. 23. Mössbauer spectra of ferrous FeS_4 complexes at 4.2 K. (**a**) $[Fe(SPh)_4]^{2-}$ in a transverse field of 2.6 T, (**b**) reduced rubredoxin in a parallel field of 2.4 T and (**c**) $[Fe(dts)_2]^{2-}$ in a transverse field of 2.6 T. Taken from Coucouvanis et al. (1981) J. Am. Chem. Soc. 103: 3350

Mössbauer spectra measured in strong fields (Fig. 23) are typical for mixed hyperfine interaction, i.e. the assembly of line positions shows that the nuclear Zeeman term and the electric quadrupole term are of comparable sizes. However, the lines are relatively sharp and the pattern depends little on the orientation of the applied field with respect to the γ-rays. This behavior indicates that the nuclear interactions are roughly the same for each molecule, irrespective of its orientation, and therefore it follows that the spin expectation values $\langle \vec{S} \rangle$ must have a preferred orientation in the molecular frame. This condition is met for the highest and lowest sublevels of the $S = 2$ multiplet, if the zero-field interaction has rhombicity E/D close to 1/3. For positive D the y-axis is the easy direction of magnetization, while for negative D it is the z-axis. Spin-Hamiltonian analyses for ferrous rubredoxin yielded D = 7.8 cm^{-1}, E/D = 0.28 [56] and D = -6 cm^{-1} and E/D = 0.19 for desulforedoxin [55].

4.3.2.2 Magnetic Susceptibility Experiments

Independent determinations of the fine-structure parameters D and E by magnetic susceptibility measurements usually present difficulties which are primarily due to the large number of unknown parameters. In addition to D and E/D the g-values g_x, g_y and g_z are required for calculations of χ. Determinations of these parameters by adjusting them to the experimental data is a complicated problem which requires a well-defined methodology in order to obtain reliable results [60]. Particularly difficult for powder samples is the determination of the sign of D and the value of the rhombicity parameter E/D. If the zero-field interaction is weak, like for ferric systems, the sensitivity and accuracy of the measurements often turn out to be insufficient. In the case of weak zero-field splittings experimental perturbations like weak spin coupling in non-perfectly diluted systems or traces of paramagnetic impurities may heavily interfere with the low-energy effects to be discriminated [29].

For ferrous systems, where the splittings are usually larger, the method was applied successfully, especially for a series of model complexes containing the FeS_4 unit [60]. The effective magnetic moment of 4.9β measured in the temperature range 100 K–300 K corresponds to the well isolated S = 2 ground state. In order to provide a guidance to the influence of zero-field splitting at low temperatures the magnetic field dependence of the susceptibility (isothermes) for various combinations of D and E/D was computed [60]. The main result of the search is that qualitative differentiation is achieved for the field dependence at temperatures comparable to the zero-field splitting D. Figure 24 shows the results of this computation for temperatures 4.2 K and 2 K and g = 2. The following main conclusions were drawn by comparing the different parts of the Fig. 24: (i) The differences in the field dependences for values of D of opposite sign become most pronounced as D increases and the temperature decreases. For |D| < 4 K, lower temperatures than 2 K are required in order to determine the sign of D. (ii) The influence of E/D is more pronounced at low fields and at 2 K more than at 4 K. Furthermore, the effect is stronger for D > 0 than for D < 0.

Approximate values for D were estimated in temperature-dependent measurements from the temperatures at which the effective magnetic moments approach a constant value. All the spin substates are populated at this temperature and for S = 2 the relation |4D| $\lesssim$ kT holds. Explicit spin-Hamiltonian simulations of field- and temperature-dependent magnetic susceptibility measurements yield a unique set of zero-field parameters, which were corroborated by Mössbauer spectroscopy [60].

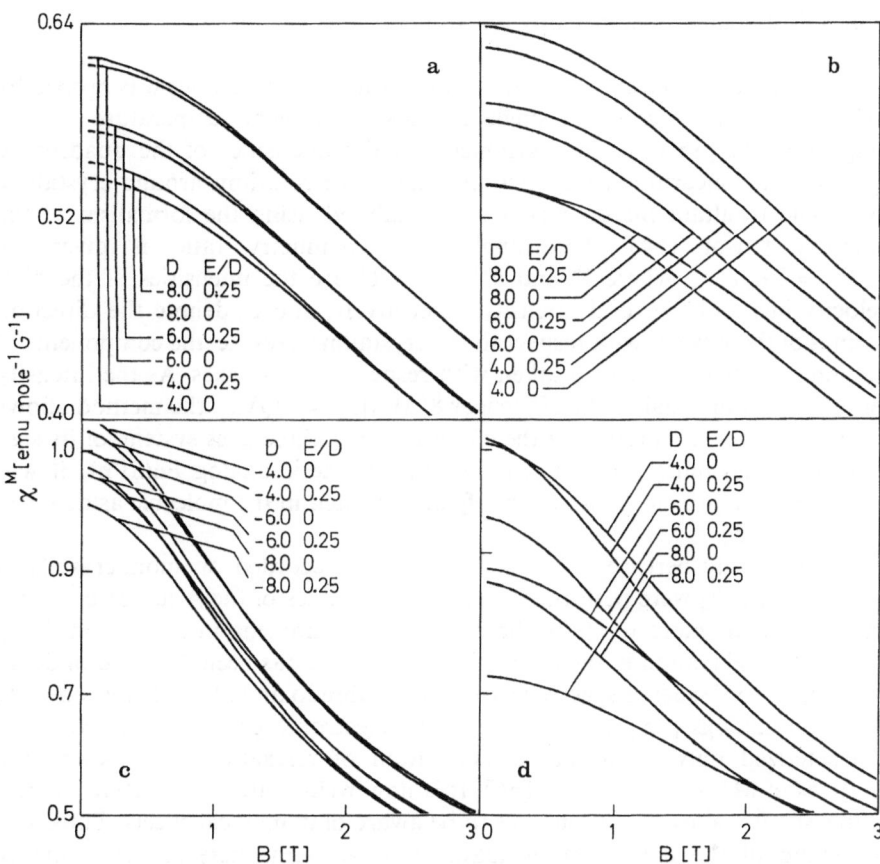

Fig. 24. Spin-Hamiltonian computations of the field dependence of the molar magnetic susceptibility χ^M for various combinations of D and E at 4.2 K (**a** and **b**) and at 2 K (**c** and **d**) for negative (**a** and **c**) and positive (**b** and **d**) D values. χ^M is given in emu mole^{-1} G^{-1}, which may be converted in SI units (m^3 mole^{-1}) by multiplying with $4\pi\ 10^{-6}$. Taken from [60]

4.3.2.3 Orientation of the Zero-Field Tensor

A variety of synthetic analogues of Fe–S centers have been prepared and studied in parallel with the proteins [59, 61] in order to elucidate the structure and electronic properties of the active centers. However, spectroscopic investigations of polycrystalline samples do not provide the orientation of the principal axes of the g-, A- and D-tensors with respect to the molecular frame. For this purpose single-crystal measurements are required. Molecular orbital calculations have also been used to study orientations of fine- and hyperfine tensors in molecules [62].

4.3.2.4 Mössbauer Spectroscopy on Monocrystals

The orientation of the efg tensor of ferrous iron in Fe–S analogs was probed by Mössbauer spectroscopy on single crystals at ambient temperature in zero magnetic field [15, 63]. The asymmetry in the intensities of the quadrupole spectrum was measured as a function of angle for rotations around crystalline axes. The resultant malus curves were analysed using the formalism of the intensity tensor [64]. Accordingly, the asymmetry ratio is given by $I^h/I^{tot} = \Sigma I_{pq}e_pe_q$, where I^h and $I^{tot} = I^h + I^l$ are the intensities of the high velocity line and the total intensity, respectively, and e_p denote the direction cosines of the γ-beam with respect to the crystalline axes, i.e. the components of the intensity tensor are also taken with respect to these axes. As the intensity tensor is proportional to the efg tensor $8|\Delta E_Q|I_{pq} = eQV_{pq}$, the method allows estimates of the orientation of the efg in the crystalline axes system, of its sign and of the asymmetry parameter. For the synthetic analog $Fe(SPh)_4$ it was found that the principal axes of the efg are oriented in the molecule as depicted in Fig. 25 [15].

Magnetically perturbed Mössbauer spectra recorded on monocrystals of ferrous $Fe(SPh)_4$ were performed to check the values of the A-tensor components and furthermore to study the relaxation phenomena in this system [15]. Spin-lattice relaxation in the high-spin ferrous state has been analyzed in detail by means of Mössbauer spectroscopy first in rubredoxin [14] and later on in its synthetic analogs [65]. In other cases the electronic ground states could be characterized only by proper consideration of relaxation processes in the analyses of Mössbauer spectra [66]. It is now well-established that at temperatures slightly above 4.2 K one has to be aware of dynamical effects. Thus, safe command of the data analysis taking into account these effects is highly desirable in order to achieve reliable spin-Hamiltonian results.

Consequently, the analysis of monocrystal spectra of $Fe(SPh)_4$, recorded in a range of temperatures and applied fields provided improved values for spin-Hamiltonian parameters, compared to previous determinations on polycrystalline material by including relaxation effects [65]. The orientation of the principal axes system of the hyperfine tensor in the molecular frame is in agreement with previous molecular-orbital calculations [62], within the limits of about 10°. Moreover, the monocrystal results verify the theoretical finding that A-, g-, and efg tensors are nearly collinear. Comparisons of fine and hyperfine parameters

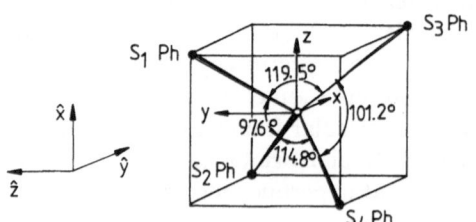

Fig. 25. Schematic structure of $[Fe(SPh)_4]^{2-}$. The axes $\hat{x}$, $\hat{y}$, $\hat{z}$ denote the principal axes system of the efg. Taken from [62]

reveal essentially identical ground states for ferrous iron in $Fe(SPh)_4$ and reduced rubredoxin [62, 15]. The observables are consistent with an orbital state of predominantly $|3z^2 - r^2\rangle$ character, as derived from the molecular orbital calculations. The example of desulforedoxin shows that this ground state by no means can be regarded as unique for the $Fe^{2+}S_4$ unit [15]. Especially the negative D value and the positive quadrupole coupling constant indicate that the dominant character of the orbital ground state in this case is most likely $|x^2 - y^2\rangle$. This shows that caution must be exercised in transferring the parameters derived from mononuclear $Fe^{2+}S_4$ units to the calculation of effective hyperfine parameters of polynuclear iron sulfur units.

4.4. Spin-Admixed Iron (III) Complexes

During the last decades, a series of systematic studies on the correlation between X-ray structure, spin state, and fine and hyperfine interactions in porphyrinatoiron(III) complexes has appeared in the literature (see e.g. Ref. [67]). Much interest was thereby focused on mixed-spin (S = 5/2–3/2) states in porphyrinatoiron(III) complexes. Detailed analyses of the magnetic and electronic properties of the mixed-spin ground states by means of various spectroscopies are found in the study of ferricytochrome c′ [68, 69], of bis(3,5-dichloropyridine)porphyrinatoiron(III) [70], and of (triflato)-(aquo)porphyrinatoiron(III) [10]. In this section we will consider the latter study in more detail and discuss the interpretation of the electronic and magnetic data in terms of the 10-state ($^6A_1$, $^4A_2$) model for spin mixture, presented in Sect. 2.3.2.

The X-band EPR spectra of the (triflato)(aquo)porphyrinatoiron(III) complex provide spectroscopic g values $g_x^{eff} \sim g_y^{eff} = 5.7$, and $g_z^{eff} = 2.0$. A comparison with the effective g values for the $|S, M_s = \pm 1/2\rangle$ Kramers doublet, deriving from the effective spin ($S^{eff} = 1/2$) Zeeman Hamiltonian $\beta\vec{B}\cdot\tilde{g}\cdot\vec{S}^{eff}$, i.e. $2S + 1$, $2S + 1$, 2, reveals a ground state of prevailing S = 5/2 character. Simulations of the EPR spectra yield values for $\Delta g_\perp = g_x^{eff} - g_y^{eff}$ which slightly depend on the value of the line width adopted in the simulations, but is less than 0.5. As already noted, splittings of the $g_\perp^{eff} = 6$ signal are caused by the rhombic term in the spin Hamiltonian. A first-order treatment of the rhombic perturbation on the tetragonal ('axial') zero-field interaction leads to $\Delta g_\perp = 48E/D$. Hence, the rhombicity parameter in the complex is small (E/D < 0.01) and implies either a small value of the ligand-field splitting w or a large value of Δ_2 in the term scheme in Fig. 12. The $g_\perp$ value in the spin (S = 5/2)-Hamiltonian description (Eq. (1)) of the ground spin multiplet is related to the effective $g_\perp$ value by $g_\perp = g_\perp^{eff}/3 = 1.9$. The latter value implies, according to eq. (20b), spin–mixture coefficients (see Eq. (18)) with $a_{1/2}^2(S = 5/2) = 0.85$ and $b_{1/2}^2$ (S = 3/2) = 0.15. Substitution of $g_\perp = 1.9$ in Eq. (20c) yields for the ratio of the ligand-field splitting and the spin–orbit coupling constant the value $\Delta_1/\zeta = 2.15$.

The temperature dependence of the intensity of the EPR signal furnishes information about the thermal depopulation of the resonant doublet, $|\pm 1/2\rangle$, by thermal population of the non-resonant doublets, $|\pm 3/2\rangle$ and $|\pm 5/2\rangle$, and thus about the zero-field splittings. However, the strong relaxational line broadening found in the complex makes the resulting values for the splittings rather inaccurate. In order to gain complementary data for the determination of the zero-field parameters, magnetic susceptibility measurements were performed in the temperature range 2–300 K. The effective magnetic moment, μ^{eff}, increases monotonically from 4β at 2 K to a value of about 5.66β at temperatures above 150 K. The high-temperature value of μ^{eff} lies in between the spin-only values 3.87β for spin S = 3/2 and 5.92β for S = 5/2, and is a measure for the spin mixture in the ground state. The slope of the $\mu^{eff}(T)$ curve in the temperature range 2–150 K depends on zero-field splitting ($\sim \zeta^2/\Delta_1$). As $\Delta_1 \sim \zeta$, the magnetic data were least-squares fitted with the theoretical curve derived from the Van Vleck formula, Eq. (41), with the 10-state model. The optimal parameter values, obtained under the restriction, $\Delta_1/\zeta = 2.15$, imposed by the $g_\perp$ value, are $\Delta_1 = 491$ cm^{-1} and $\zeta = 229$ cm^{-1}, and yield an excellent fit of the experimental data. The zero-field splittings, evaluated from Eq. (19) using the optimal parameters, are $\varepsilon_{5/2} - \varepsilon_{3/2} = 74$ cm^{-1} and $\varepsilon_{3/2} - \varepsilon_{1/2} = 31$ cm^{-1} (D $\sim (\varepsilon_{3/2} - \varepsilon_{1/2})/2 = 16$ cm^{-1}).

Complementary to the EPR and susceptibility analyses, Mössbauer measurements were performed for the determination of the electric and magnetic hyperfine parameters. The Mössbauer spectra recorded in a crystalline powder without applied field exhibit a doublet with quadrupole splitting $\Delta E_Q = 2.2$ mms^{-1} and isomer shift $\delta(\alpha\text{-Fe}) = 0.43$ mms^{-1}, nearly independent of the temperature. The low-temperature spectra can be resolved into six-line patterns by applying a magnetic field. Spin-Hamiltonian analysis of the magnetic splittings provides information about the magnetic hyperfine fields and yields the sign of the main component of the electric field gradient tensor at the iron nucleus ($\Delta E_Q = +2.2$ mms^{-1}). The 4.2 K spectra originate mainly from the lowest Kramers doublet $|\pm 1/2\rangle$, because the zero-field splittings are large compared to the thermal energy (D $\gg$ kT). The complex then possesses an easy plane of magnetization, giving spin expectations values mainly within the molecular xy plane for the majority of the crystal orientations relative to the applied field. As noted before, the analysis then furnishes accurate values for the x and y components of the $\tilde{A}$ tensor in the ground doublet, $A_\perp/\beta_n g_n = -17$ T, but only a rough estimate for the z component, $A_{||}/\beta_n g_n = -13.5$ to -17.0 T. The anisotropy of the A tensor is, within the tolerance, compatible with the dipolar field predicted for the admixture of 15% $^4A_2$ character into the ground doublet, deduced from the $g_\perp$ value. Field-dependent measurements at 4.2 K show that the hyperfine fields, and the associated line splittings in the spectra, reach their saturation value already for small applied fields (B $\gtrsim$ 10 mT). A further increase of the applied field up to 6 T leaves the hyperfine field almost constant, because the Zeeman interactions between the Kramers doublets are

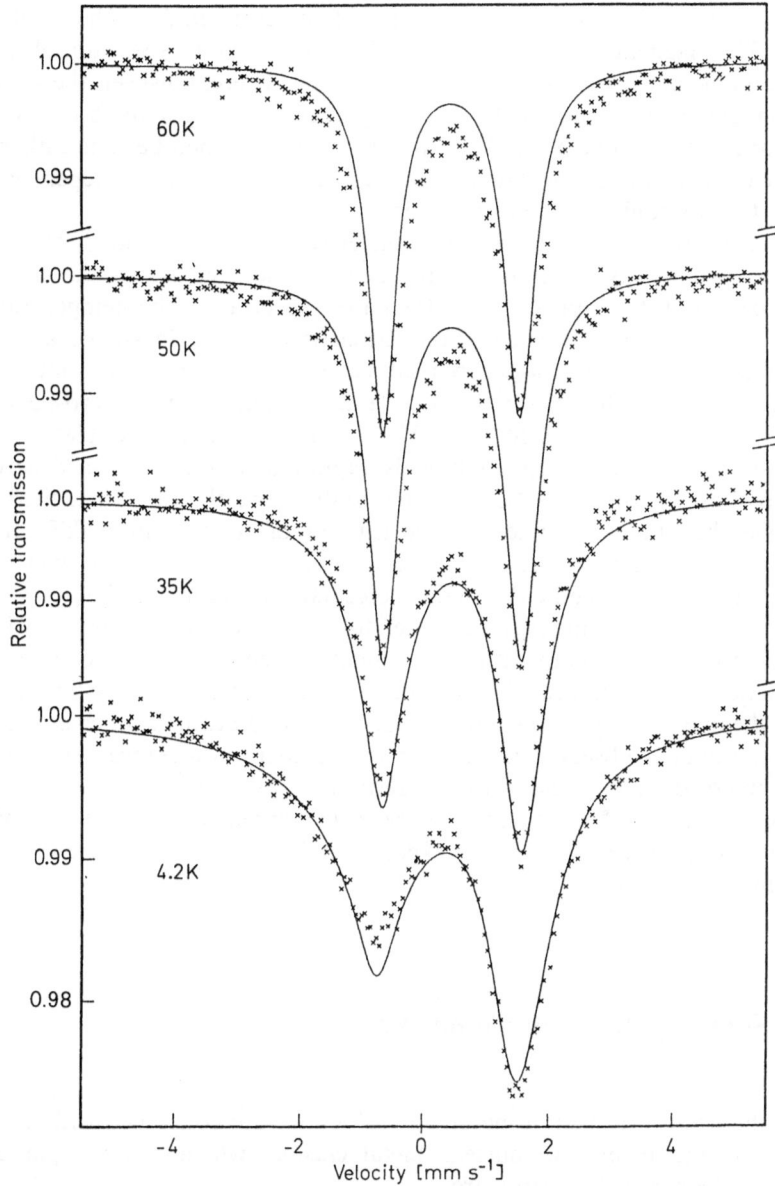

Fig. 26. Mössbauer spectra of a (triflato)(aquo)porphyrinato-iron(III) complex recorded at temperatures ranging from 4.2 K to 60 K in an applied field of 12 mT perpendicular to the γ rays. The *solid lines* represent the simulations based on spin (S = 5/2) Hamiltonian, Eq. (1), taking into account spin–spin relaxation with a relaxation constant $w^{ss} = 4.4 \times 10^7$ rad s^{-1} for the 4.2 K spectrum and spin–lattice relaxation with $w^{sl} = 9 \times 10^3$ rad s^{-1} K^{-3} for the other spectra. Parameters used: $\delta(\alpha\text{-Fe}) = 0.43$ mms^{-1}, $\Delta E_Q = +2.2$ mms^{-1}, $\eta = 0$, $g_\perp = 1.9$, $g_{||} = 2.0$, $A_{||}/g_n\beta_n = -13.5$ T, $A_\perp/g_n\beta_n = -17$ T, $E = 0$ cm^{-1}, and $D = 15.6$ cm^{-1}

still small compared with the zero-field splitting ($D \gg \beta B$). The zero-field splitting could therefore not be accurately determined from the 4.2 K spectra.

The magnetic properties of the excited $|\pm 3/2\rangle$ and $|\pm 5/2\rangle$ doublets are only accessible at elevated temperatures ($kT > D$). There, however, the slow relaxation behaviour, found below 10 K, gradually changes to fast relaxation at temperatures above 50 K. The Mössbauer spectra then become rather insensitive to variations in the $\tilde{A}$-tensor components, so that the M_s-dependence of the $\tilde{A}_M$ tensors could not be established.

The value of the zero-field parameter D of around 16 cm^{-1}, obtained from the susceptibility data, was confirmed by the analysis of the intermediate relaxation behaviour of the Mössbauer spectra in the temperature range 10–50 K. The analysis is based on stochastic theory of line-shape, and takes the effects of spin–spin and spin–lattice relaxation explicitly into account. Favourable conditions for the precise determination of the zero-field splitting are found at weak applied fields. These split the Kramers doublets only a little without mixing them. In small fields, rapid relaxation is expected between the two substates of the $|\pm 1/2\rangle$ ground doublet and the $|\pm 3/2\rangle$ excited doublet, while the $|\pm 5/2\rangle$ doublet can be left out of consideration ($kT < 6D$). The resulting magnetic hyperfine field generated by the $|\pm 1/2\rangle$ doublet points, as stated earlier, perpendicular to the molecular z axis, whereas the $|\pm 3/2\rangle$ doublet gives rise to a component parallel to the z axis, which grows with rising temperature. The temperature-dependent Mössbauer spectra consist therefore of a quadrupole split resonance with one or the other of the two lines broadened by the magnetic hyperfine interaction. The temperature at which equal intensity of the lines is observed is sensitive to the energy difference between the two Kramers doublets, and thus on zero-field splitting. In Fig. 26, Mössbauer spectra taken at various temperatures in an applied field of 12 mT are displayed. The overall agreement with the experimental spectra confirms the validity of the value $D \approx 16$ cm^{-1}.

5 Spin-Coupled Iron Complexes

In the field of transition metals in biology we are concerned with a variety of homo- and hetero-polynuclear metal clusters which exhibit spin coupling. Examples for such clusters are

(i) the large family of molecules which is based on the common building block Fe_2S_2 occurring in iron-sulfur proteins and synthetic analogs, and

(ii) the metal clusters which have bridges other than thiolates or sulfides, for example oxygen in Cu ... Cu (hemocyanin), in Fe ... Cu (cytochrome oxidase), in Fe ... Fe (hemerythrin), in Mn ... Mn (photosystem II), in $(Fe ... Fe)_n$ ferritin.

A further example for spin coupling is provided by mononuclear transition-metal complexes, with the paramagnetic metal site and an additional non-metallic paramagnetic site forming a spin-coupled system. Representatives of this category are the iron(IV)-porphyrin cation radical interaction in horse-radish peroxidase I (HRP I), catalase, chloroperoxidase I (CPO I) and related synthetic analogs, and the iron(II)-dioxygen interaction in iron-substituted horse liver alcohol dehydrogenase.

The specific spin-coupling properties of such systems together with their structural diversity provides a wealth of objects for the study of magneto-structural correlations. Therefore these systems are excellent targets for the spectroscopic study of structural dependencies of

  (i) localized and delocalized mixed-valence states
 (ii) electron-hopping processes between metal sites
(iii) spin coupling containing Heisenberg-exchange, spin-dipolar and double-exchange interactions.

5.1 Extension of the Spin Hamiltonian

The spin Hamiltonian, which contains the relevant contributions for a spin-coupled system, consisting of two paramagnetic sites ($\vec{S}_1$ and $\vec{S}_2$), reads

$$\mathscr{H}_e = \mathscr{H}_1 + \mathscr{H}_2 + \mathscr{H}_{12}, \tag{47}$$

with $\mathscr{H}_1$ and $\mathscr{H}_2$ describing the site-specific properties and with the interaction term $\mathscr{H}_{12}$ defined as

$$\mathscr{H}_{12} = - \vec{S}_1 \cdot \tilde{J} \cdot \vec{S}_2 + T. \tag{48}$$

There are two major contributions to the spin-coupling tensor, i.e.: $\tilde{J} = J\tilde{I} + \tilde{J}^d$.

The first contribution results from the Heisenberg-exchange interaction, which is expressed by an isotropic term, $- J\vec{S}_1 \cdot \vec{S}_2$, with $\vec{S}_1$ and $\vec{S}_2$ being the spin operators of the two paramagnetic sites. In the major part of the complexes, the two paramagnetic sites are in definite valence states and therefore possess good local-spin quantum numbers. Due to the fact that normally the paramagnetic sites are separated by one or more (diamagnetic) ligand bridges, this coupling mechanism is called indirect exchange or superexchange. The exchange-coupling constant J is, because of the indirect character of the coupling, generally small on the chemical energy scale, i.e. $|J| < 5 \times 10^2$ cm^{-1}.

The second contribution to $\tilde{J}$ arises from the magnetic dipolar interaction

$$\mathscr{H}^d = [\vec{M}_1 \cdot \vec{M}_2 - 3(\vec{M}_1 \cdot \vec{r})(\vec{M}_2 \cdot \vec{r})/r^2]/r^3 \tag{49}$$

between the two dipole moments $\vec{M}_1$ and $\vec{M}_2$ at distance $\vec{r}$ which, when integrated over the whole space, can be written as $\vec{S}_1 \cdot \tilde{J}^d \cdot \vec{S}_2$. The contribution $\tilde{J}^d$ to $\tilde{J}$ is traceless ($J_x^d + J_y^d + J_z^d = 0$) [71], with values $|J_i^d|$ (i = x, y, z) for the systems under study here being in the order of < 0.5 cm^{-1} due to their r^{-3}-dependence.

The remaining term in Eq. (48) represents the double-exchange interaction which occurs in delocalized mixed-valence complexes. The double-exchange mechanism was originally proposed by Zener [72], as explanation for the ferromagnetism observed in a number of mixed-valence magnetites of perovskite structure [73]. In the subsequent analysis of the double-exchange mechanism in dimers by Anderson and Hasegawa [74], the interaction was formulated in terms of a simple spin Hamiltonian, possessing two equidistant energy subspectra $E_\pm(S_t) = \pm B(S_t + \frac{1}{2})$, with S_t denoting the system-spin quantum number of the dimer and B the double-exchange parameter. The double-exchange mechanism can be illustrated qualitatively in a symmetric dimer Fe ... Fe by considering the delocalization properties of an excess d-electron of β spin which is added to $Fe_{(1)}(3d\alpha)^5 \ldots Fe_{(2)}(3d\alpha)^5$. In the "antiferromagnetic" configuration $Fe_{(1)}(3d\alpha)^5 \ldots Fe_{(2)}(3d\beta)^5$ the excess electron has to be localized at site (1) because of Pauli exclusion. In the "ferromagnetic" configuration, $Fe_{(1)}(3d\alpha)^5 \ldots Fe_{(2)}(3d\alpha)^5$, however, the excess electron may be delocalized over sites (1) and (2) in gerade or ungerade orbital states, with a gain or a loss of resonance energy, leading to an energetic lowering of the "ferromagnetic" ground state relative to the "antiferromagnetic" state. In delocalized-valence dimers with a super-exchange term favoring antiparallel spin coupling ($J < 0$), the actual ground-state spin is dependent on the relative magnitude of the super-exchange parameter J *and* of the double-exchange parameter B.

In the following we will discuss various spin-coupled systems which are examples for the relative importance of the different contributions to the spin Hamiltonian.

5.2   Iron(III)-Iron(III) Dimers

5.2.1   Antiparallel Spin-Coupling

In the oxidized state the $[Fe_2S_2]^{2+}$ cluster in ferredoxin proteins and synthetic analogs exhibits a Mössbauer spectrum [75] (Fig. 27a) with one quadrupole doublet or two slightly different doublets, depending on whether the terminal iron ligands are equivalent (i.e. cysteine in *Azotobacter* protein, adrenodoxin, spinach ferredoxin, parsely ferredoxin etc.) [76] or inequivalent (i.e. cysteine and probably histidine in the Rieske protein) [53]. The values of the quadrupole splitting ($\Delta E_Q \sim 0.7$ mms^{-1}, independent of temperature below 250 K) and of the isomer shift at 4.2 K, $\delta(\alpha$-Fe$) \sim 0.2$ mms^{-1}, are characteristic for the ferric high-spin state of both iron sites, tetracoordinated by sulfur ligands. In the Rieske protein one of the two sites exhibits $\Delta E_Q \sim 0.9$ mms^{-1} and $\delta(\alpha$-Fe$) \sim 0.3$ mms^{-1} [53].

The application of an external field perpendicular to the γ-beam at 4.2 K broadens the Mössbauer spectrum [75] (Fig. 27b). The analysis of this spectrum reveals that the broadening is solely due to nuclear Zeeman interaction, but there is no additional magnetic hyperfine contribution due to unpaired Fe 3d

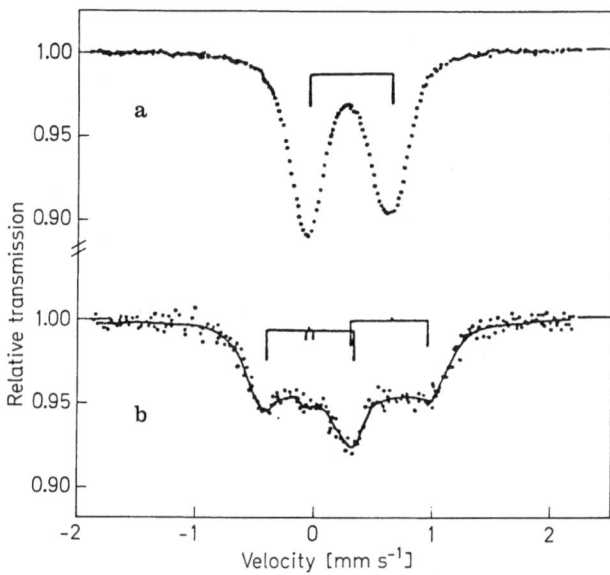

Fig. 27. Experimental Mössbauer spectra of *Euglena* ferredoxin (**a**) at 4.2 K in zero-applied field and (**b**) in 3 T-applied field, perpendicular to the γ-beam. The solid line represents a simulation with zero-magnetic hyperfine contribution; i.e. the magnetic splitting arises solely from the applied field (see stick-spectrum). Taken from [75]

electrons. The iron in $[Fe_2S_2]^{2+}$ is therefore in a non-magnetic (diamagnetic) state (see left side of Fig. 28) at this temperature. This is also the reason why the system is EPR-inactive.

In this system the exchange term $-J\vec{S}_1\cdot\vec{S}_2$ dominates all other terms in the spin Hamiltonian. Hence, the sequence of energy states for $[Fe_2S_2]^{2+}$ with $J < 0$ is reasonably represented by Fig. 29a. This situation simplifies the analysis of the temperature-dependence of the molar susceptibility (in SI-units) of $[Fe_2S_2]^{2+}$ [77] by

$$\chi^M_{Fe\ldots Fe} = \mu_0 \frac{N_A g^2 \beta^2}{3kT} \frac{\sum\limits_{S_t=(S_1-S_2)}^{(S_1+S_2)} S_t(S_t+1)(2S_t+1)\exp\dfrac{JS_t(S_t+1)}{2kT}}{\sum\limits_{S_t=(S_1-S_2)}^{(S_1+S_2)} (2S_t+1)\exp\dfrac{JS_t(S_t+1)}{2kT}}, \quad (50)$$

which is a consequence of Eq. (34) and Eq. (41), and which provides a mean for deriving the exchange-coupling constant J. An example for this procedure is given in Fig. 29b [78]. The measured magnetic moments in the given temperature range are relatively small (because |J| is relatively large) so that background contributions to the total moment may be significant. These contributions (paramagnetic impurities, diamagnetic contributions from the sample etc.) are common to all static magnetic measurements of dilute paramagnetic ions in large molecules and may become even dominant particularly for systems with

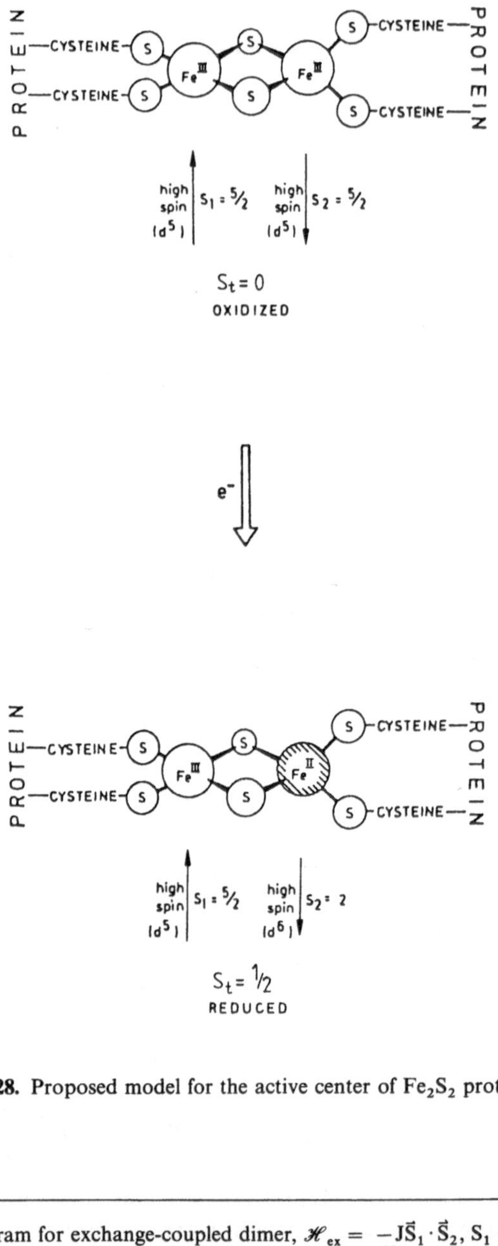

Fig. 28. Proposed model for the active center of Fe_2S_2 proteins

Fig. 29a. Energy diagram for exchange-coupled dimer, $\mathscr{H}_{ex} = -J\vec{S}_1 \cdot \vec{S}_2$, $S_1 = S_2 = 5/2$, $J < 0$. **(b)** Temperature dependence of the molar susceptibility per iron, χ_{Fe}^M, of $(Ph_4As)_2[Fe_2S_2(S_2\text{-}o\text{-xyl})_2]$ and calculated curve for $J = -296\,cm^{-1}$. Taken from [78]. Note that χ_{Fe}^M is given in emu mole^{-1} G^{-1}, which may be transferred into SI-units (m^3 mole^{-1}) by multiplying with $4\pi \times 10^{-6}$ Effective magnetic moments per iron, μ^{eff}, at various temperatures, are calculated from χ_{Fe}^M (in emu mole^{-1} G^{-1}) using $\mu^{eff} = \sqrt{8\chi_{Fe}^M T}\,\beta$ and from χ_{Fe}^M (in SI-units) using $\mu^{eff} = 797.5\sqrt{\chi_{Fe}^M T}\,\beta$. **(c)** Dependence of exchange coupling in Fe(III)–Fe(III) dimers from the type of bridges; (*1*): $[Fe_2(edt)_4]^{2-}$ from Herskovitz et al. (1973) Inorg. Chem. 12: 249; (*2*): $[Fe(salen)]_2S$ from Mitchell et al. (1973) J. Inorg. Nucl. Chem. 35: 1385; (*3*): $[Fe_2S_2(S_2\text{-}o\text{-xyl})_2]^{2-}$ from [78]; (*4*): spinach ferredoxin from Palmer et al. (1971) Biochim. Biophys. Acta 245: 201

singlet ground states at low temperatures. It is therefore important to correct the measured data for these contributions prior to the analysis described above.

A general feature concerning the magneto-structural correlation in sulfur-bridged Fe(III)–Fe(III) dimers, derived from magnetic susceptibility measurements and structural studies, was reported to be that the exchange coupling

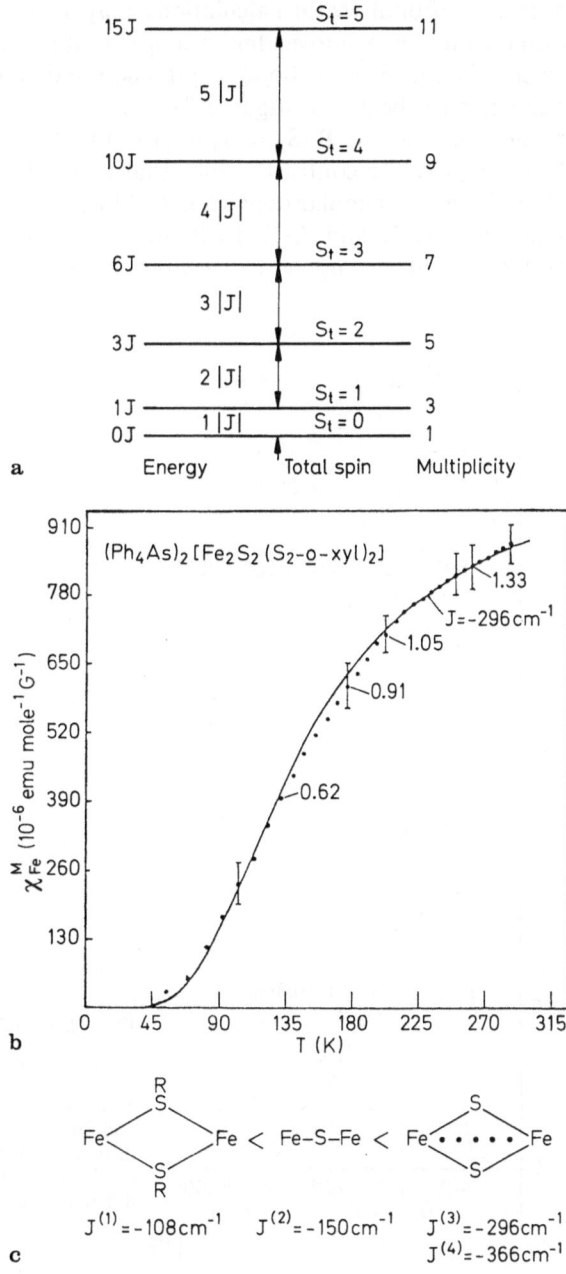

increases as Fe . . . Fe distances decrease and thiolate bridges are replaced by sulfide bridges (Fig. 29c) [78].

5.2.2 MO Results and the "Broken Spin-Symmetry" Concept

Molecular orbital (MO) calculations may provide an estimate of exchange coupling for a given molecular structure. Using a simple coupled-cluster model, structural dependencies (bond length and bond angle) of the exchange-coupling constant have been investigated for a double-bridged cluster with two paramagnetic sites A and B ($S_A = S_B = \frac{1}{2}$) and with two diamagnetic bridges C and D (Fig. 30) [79]. In contrast to the usual suggestions in the literature [80, 81], it is found that the angular dependence of the exchange-coupling constant on the angle Θ (A–C–D and A–D–B) is not necessarily associated with directional properties of atomic np, nd . . . orbitals. These model calculations [79] support

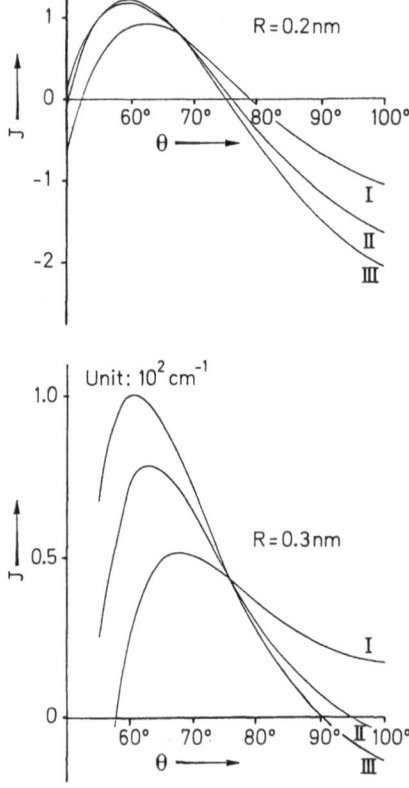

Fig. 30. Coupled-cluster calculations of indirect exchange for the double-bridged system A⟨^C_D⟩B with paramagnetic sites A and B ($S_A = S_B = 1/2$) and diamagnetic ligands C and D, depending on distance R between A and B and on angle Θ (A–C–B and A–D–B). I, II, and III refer to three different calculational procedures. Taken from [79]

earlier results [82, 83] in that the experimental trends can already be explained on the basis of spherically symmetric atomic orbitals.

MO calculations, based on the LCAO-Xα-method, were performed on $[Fe_2S_2(SR)_4]^{2-}$ with R = H, CH_3, $(C_6H_4(CH_2)_2)_{1/2}$, which are obviously more realistic models for the Fe(III)–Fe(III) dimers in ferredoxin proteins [84–86]. The calculations were performed in the high-spin ("ferromagnetic") state, i.e. with 10 (unpaired) α electrons, and in the low-spin ("antiferromagnetic") state, possessing an equal number of α and β electrons. These computations result in two paramagnetic states of spin 5/2 mainly localized at the two iron sites, which are oriented parallel and anti-parallel for the "ferromagnetic" and "antiferro-magnetic" states, respectively. The "antiferromagnetic" state, thus obtained, is not a pure spin singlet, and has therefore been denoted by Noodleman [87] as the "broken spin-symmetry" state. The calculations successfully describe charge distributions as derived from comparing calculated and measured quadrupole splittings ΔE_Q and asymmetry parameters η. Moreover, the exchange-coupling constant J has been derived qualitatively correct (i.e. correct sign and correct order of magnitude [84–86]) from the difference between the total energies of the "broken spin-symmetry" state and the high-spin state.

5.3 Iron (III)-Iron (II) Dimers

5.3.1 Antiparallel Spin Coupling

The proposed model for the one-electron reduction of the $[Fe_2S_2]^{2+}$ core of iron-sulfur proteins to $[Fe_2S_2]^{1+}$ is shown in Fig. 28. Structural details of binuclear clusters of this type (bond length, bond angles, bridging ligands, terminal ligands) are believed to tune the redox potential for this reaction. We have indicated in the last section also that magnetic properties are influenced by the structural details of these clusters. Therefore the detection and understanding of magnetic properties provides a mean for the indirect sensing of structural changes and reaction pathways.

Experimental J values for $[Fe_2S_2]^{1+}$ are around -200 cm^{-1} as derived for ferredoxins of blue green algae [88] and spinach [89] via magnetic susceptibility measurements and data analyses as described for $[Fe_2S_2]^{2+}$. The decrease of $|J|$ values for the reduced clusters compared to the $|J|$ values for the oxidized clusters (Fig. 29c) correlates qualitatively with the increase of Fe . . . Fe separation for $[Fe_2S_2]^{1+}$ by ~ 0.1 Å compared to $[Fe_2S_2]^{2+}$ [90]. Moreover the electron added by the reduction may cause (i) a lowering of the energy of the high-spin states relative to the low-spin state due to partial delocalization of this electron (see Sect. 5.3.3.2), and/or (ii) a change in J due to the reduction in the number of exchange-coupled electron pairs.

A consequence of the antiparallel spin coupling in $[Fe_2S_2]^{1+}$ is that the total spin becomes $S_t = 1/2$ as indicated in Fig. 28 (right side). The existence of $S_t = 1/2$ is demonstrated by EPR measurements which, for a variety of proteins containing the $[Fe_2S_2]^{1+}$ core, yield effective g-values $g'_z \sim 2$ and $g'_{x,y} < 2$ [91]. It is straightforward with the spin-coupling model [92] to show that these effective g-values indeed reflect the proposed coupling scheme. The effective g-tensor of the coupled system is given by

$$\tilde{g}' = \frac{\vec{S}_1 \cdot \vec{S}_t}{S_t^2} \tilde{g}_1 + \frac{\vec{S}_2 \cdot \vec{S}_t}{S_t^2} \tilde{g}_2, \tag{51}$$

with the factors

$$\frac{\vec{S}_{1(2)} \cdot \vec{S}_t}{S_t^2} = \frac{S_t(S_t + 1) + S_{1(2)}(S_{1(2)} + 1) - S_{2(1)}(S_{2(1)} + 1)}{2S_t(S_t + 1)} \tag{52}$$

describing the projections of the spins $\vec{S}_1$ and $\vec{S}_2$ of the individual ions on the resultant spin direction $\vec{S}_t$. For $S_t = 1/2$, $S_1 = 5/2$ and $S_2 = 2$ it follows $\tilde{g}' = (7/3)\tilde{g}_1 - (4/3)\tilde{g}_2$. While $\tilde{g}_1$ is expected to be isotropic and equal to 2, since the Fe(III) ion is in a spherically symmetrical S-state, $\tilde{g}_2$ of the Fe(II) ion would in general be anisotropic. For a d orbital, theory predicts that $g_z \sim 2$ and $g_{x,y} > 2$ [7]. Hence, in the exchange-coupled state, the effective g-values are $g'_z \sim 2$, and $g'_{x,y} < 2$.

It was shown that there are systematic differences between the various ferredoxins and synthetic analogs arising for the average effective g-values, $g'_{av} = (g'_x + g'_y + g'_z)/3$ [93]. Most of the Fe_2S_2 proteins [91] exhibit $g'_{av} \sim 1.96$, while the Rieske center [53] and most of the synthetic analogs [93] are characterized by $g'_{av} \sim 1.91$.

There is a class of other localized mixed-valence Fe(III)-Fe(II) proteins and synthetic analogs which exhibit $g'_{av} < 2$. These are the semimet forms of hemerythrin [94], the reduced forms of the purple acid phosphatase [95], methane monoxygenase [96], pink uteroferrin [97] and several μ-hydroxo- or μ-oxo-bridged model compounds [98–101], all of them yielding EPR signals with $g'_{av} \sim 1.7$–1.8. These g'_{av} values are also believed to indicate, according to the spin-coupling model described above, the existence of antiferromagnetic superexchange interaction between high-spin Fe(III) and high-spin Fe(II) in these systems. It should be emphasized, however, that the smallness of g'_{av} in this case cannot be estimated quantitatively on the basis of this simple model, mainly because the exchange-coupling constant $|J|$ is of the same order as the local zero-field splitting D of Fe(II). In other words, the total spin $S_t = 1/2$ is no longer a good quantum number. A similar situation has been observed in the spin-coupled Fe(IV)-porphyrin cation radical system of chloroperoxidase compound I [102], where the ferryl iron ($S_1 = 1$) couples antiparallel with the radical spin ($S_2 = 1/2$) to the total spin $S_t = 1/2$, with $D_1 = 37 \text{ cm}^{-1}$ and $|J|/D_1 \sim 1$. The observed EPR data for this system are $g'_{||} = 2$ and $g'_{\perp} = 1.73$. On the basis of a spin-Hamiltonian analysis, taking full account of zero-field splitting and exchange interaction, it was shown that the effective g-values of the spin-coupled system significantly vary with $|J|/D$, reaching g'_{av}-values as small as 0.7–1.8 in the interval $0.1 < |J|/D < 1$ [102].

We now turn to the discussion of Mössbauer spectra. A consequence of antiparallel spin coupling in $[Fe_2S_2]^{1+}$ is that the valencies in this cluster are localized. This is demonstrated by the Mössbauer investigation of, for example, reduced spinach ferredoxin [76] (Fig. 31a). The spectrum recorded at 256 K consists of two quadrupole doublets with 1:1 ratio, the one with $\delta(\alpha\text{-Fe})$ = 0.29 mms^{-1}, ΔE_Q = 0.64 mms^{-1} being characteristic for high-spin Fe(III), and the other with $\delta(\alpha\text{-Fe})$ = 0.56 mms^{-1}, ΔE_Q = 2.63 mms^{-1} characteristic for high-spin Fe(II), both tetracoordinated by sulfur ligands. Thus, at least within

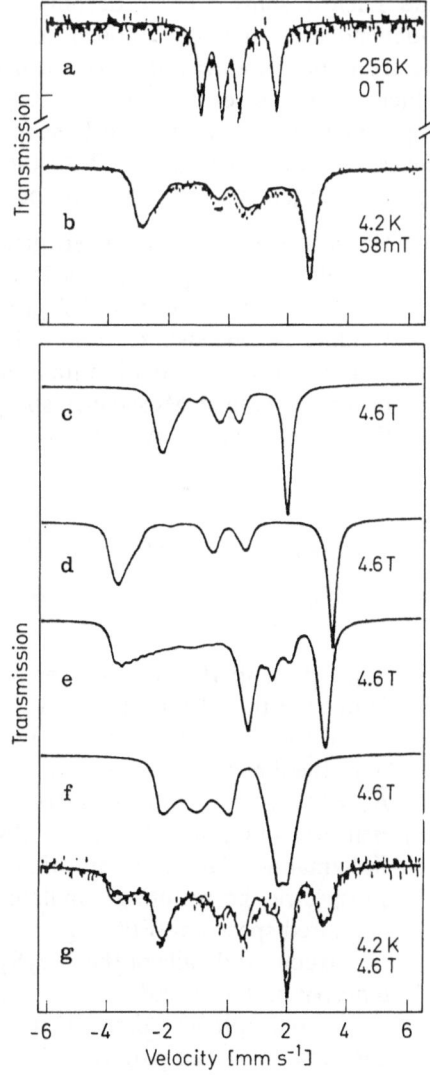

Fig. 31. Experimental Mössbauer spectra of reduced spinach ferredoxin. *Solid lines* are simulations, i.e. (c) for Fe(III) $S_z = -1/2$, (d) for Fe(III) $S_z = +1/2$, (e) for Fe(II) $S_z = -1/2$, (f) for Fe(II) $S_z = +1/2$, (g) Boltzmann-weighted sum of (c), (d), (e) and (f), superimposed over experimental spectrum. Taken from [76]. (Zero-velocity given relative to Pt-source matrix; transformation of $\delta(Pt)$ to $\delta(\alpha\text{-Fe})$ is provided by adding 0.349 mms^{-1} to $\delta(Pt)$)

the time window of $\sim 10^{-7}$ s for ^{57}Fe-Mössbauer spectroscopy, the excess-electron which reduces $[Fe_2S_2]^{2+}$ to $[Fe_2S_2]^{1+}$, is trapped at one of the two metal sites. Decreasing the temperature (in the absence of an external field) to 4.2 K and recording a Mössbauer spectrum yields a broad unresolved magnetic hyperfine pattern (not shown here). This observation may be explained in terms of coupled electron spin S and nuclear spin I yielding a resultant total angular momentum F = S + I. Solving the appropriate electron-nuclear Hamiltonian for the energy states and applying the selection rules of magnetic dipole radiation for the γ-transitions, the predicted Mössbauer spectrum has the appearance of this broad unresolved hyperfine pattern. When a field is applied in the order of the EPR-hyperfine splitting of ^{57}Fe, in practice > 2 mT, electron and nuclear spins become decoupled from each other and, instead, precess independently about the applied field at the nucleus. With this condition the electron spin produces an effective field $\vec{B} = \langle\vec{S}\rangle\tilde{A}/g_n\beta_n$ at the iron nucleus, which yields the resolved nuclear Zeeman splitting shown in Fig. 31b. If the applied field is increased to 4.6 T the measured magnetic hyperfine pattern takes the form drawn in Fig. 31g. With the spectroscopic properties of $[Fe_2S_2]^{1+}$ collected so far this spectrum may be analyzed as follows:

(i) For a $S_t = 1/2$ system the zero-field terms in the spin Hamiltonian vanish. Therefore in strong antiparallel spin coupling, $kT \ll |J|$, the spin Hamiltonian restricted to the $S_t = 1/2$ groundstate, reduces to the Zeeman term, yielding two eigenstates $|j\rangle$, j = 1, 2.

(ii) Within the slow-relaxation limit at low temperature (4.2 K) four separate, powder-averaged Mössbauer spectra have to be calculated according to the nuclear Hamiltonian for a ≡ Fe(III) and b ≡ Fe(II) and for $S_t = 1/2$, respectively:

$$\mathscr{H}_{n,a} = \mathscr{H}_{Q,a} + (7/3)\langle\vec{S}\rangle_j \cdot \tilde{A}_a \cdot \vec{I} - g_n\beta_n\vec{B}\cdot\vec{I},$$
$$\mathscr{H}_{n,b} = \mathscr{H}_{Q,b} - (4/3)\langle\vec{S}\rangle_j \cdot \tilde{A}_b \cdot \vec{I} - g_n\beta_n\vec{B}\cdot\vec{I}. \tag{53}$$

$\mathscr{H}_Q$ represents the quadrupolar part of the nuclear Hamiltonian, Eq. (21). Similar to the effective g-tensor the effective hyperfine-coupling tensors $\tilde{A}'$ of the spin-coupled system result from projections of the single-site spins onto the direction of $\vec{S}_t$. Hence, the effective nuclear Hamiltonians contain $\tilde{A}'_a = (7/3)\tilde{A}_a$ and $\tilde{A}'_b = -(4/3)\tilde{A}_b$.

(iii) Before adding up the four subspectra (Fig. 31c–f) they have to be Boltzmann-weighted appropriately. After summation of the weighted subspectra the resulting simulated spectrum then is compared with the measured spectrum (Fig. 31g).

Structural details of the $[Fe_2S_2]^{1+}$ core within different proteins (*Azotobacter* protein I and II, *Clostridium* protein, adrenodoxin, parsley ferredoxin and spinach ferredoxin) are reflected by the variations of their magnetic hyperfine pattern [76].

5.3.2 MO Results for Antiparallel Spin-Coupled Iron(III)–Iron(II) Dimers

Electronic structure calculations provide a tool to correlate molecular structures with spectroscopic properties. Applications in this area are molecular orbital calculations, based on the LCAO-Xα theory, which have been used to study

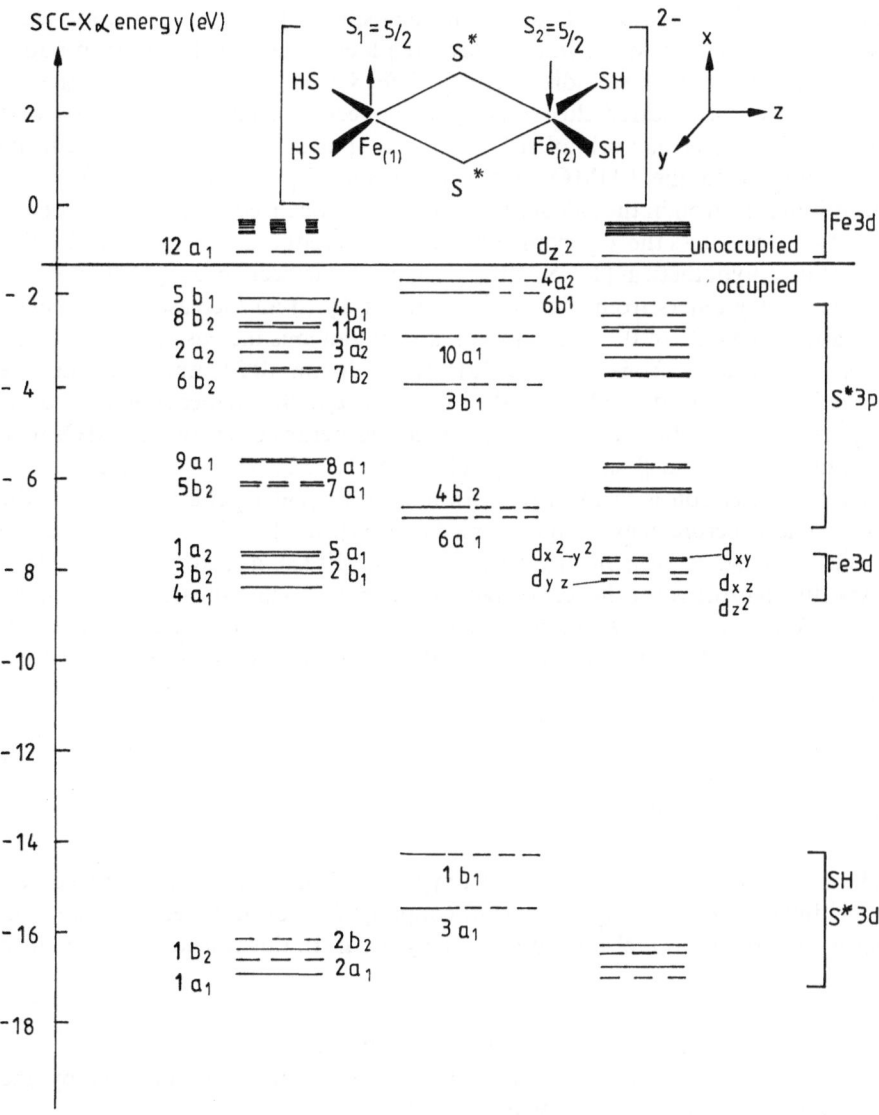

Fig. 32. One-electron MO energies of $[Fe_2S_2(SH)_4]^{2-}$ as derived from the spin-unrestricted SCC-Xα method [85]. *Full lines* correspond to α spin ("spin up") and *dashed lines* to β spin ("spin down")

binuclear iron–sulfur clusters [84–86]. In the following, as an example for these studies, we report our own investigation of synthetic analogs for reduced Fe_2S_2 ferredoxins.

The observed Mössbauer spectra of the reduced spinach ferredoxin exhibit a 1:1 superposition of ferric and ferrous high-spin spectra (Fig. 31a). Electronic structure calculations performed for the high-spin configuration ($S_t = 9/2$) of a corresponding model cluster $[Fe_2S_2(SH)_4]^{3-}$ result in delocalized MOs and, thus, lead to identical Mössbauer parameters for both iron, in disagreement with experiment. On the other hand, the "broken spin-symmetry" description leads to a set of localized MOs, which allows the localization of the electron added upon reduction of the oxidized cluster $[Fe_2S_2(SH)_4]^{2-}$. Because of the C_2 symmetry of the oxidized cluster along the molecular x axis (Fig. 32), the lowest unoccupied molecular orbital (LUMO) of α spin at site $Fe_{(1)}$ is energetically degenerate with the LUMO of β spin at site $Fe_{(2)}$. In order to force the additional electron in the calculation to occupy a localized iron orbital at one of the two iron sites, the C_2 symmetry has to be reduced. As a model for this situation a hypothetical $[Fe_2S_2(SR)_4]^{3-}$ cluster has been investigated, of which the geometry differs from the oxidized cluster by (i) an increase in the Fe–Fe distance from 0.27 to 0.275 nm and (ii) a simultaneous increase of the $Fe_{(2)}$–S* ($Fe_{(2)}$–S) distance from 0.221 (0.231) to 0.226 nm (0.236 nm), keeping the S–$Fe_{(2)}$–S angle fixed and leaving the $Fe_{(1)}$–S* ($Fe_{(1)}$–S) distances unchanged. As expected, the structural changes remove the degeneracy of the two LUMOs with α spin on $Fe_{(1)}$ and β spin on $Fe_{(2)}$ ($12a_1$ in Fig. 32), with the result that the additional electron is in the MO with mainly $Fe_{(2)}3d_{z^2, \alpha}$ character. From the electronic structure of exchange-coupled $[Fe_2S_2(SR)_4]^{3-}$ clusters ($R = H, CH_3$) electron densities $\rho(0)$, electric field gradients (EFG), g factors and magnetic hyperfine tensors $\tilde{A}$ for the ferric high-spin iron ($Fe_{(1)}$) and the ferrous high-spin iron ($Fe_{(2)}$) have been evaluated [103]. In Table 2 the calculated values are compared with the data obtained from the analyses of Mössbauer and of EPR measurements of reduced Fe_2S_2 ferredoxins and synthetic analogs.

5.3.3 Parallel Spin Coupling

The majority of known sulfur- and oxygen-bridged Fe(III)–Fe(II) mixed-valence dimers exhibits antiparallel spin coupling. However, under specific structural conditions parallel spin coupling may occur. In this respect we distinguish two cases, i.e.

(i) the exchange-coupling constant J is positive, and
(ii) the exchange-coupling constant J is negative but is dominated by the double-exchange parameter B.

In the following we give examples for these two cases.

Table 2. Calculated quadrupole splittings ΔE_Q, electric-field gradients (sign and orientation of the main component V_{zz}), asymmetry parameters η, changes in isomer shift $\Delta\delta$, effective g values and hyperfine-coupling tensors $\tilde{A}'$, as derived from spin-unrestricted SCC-Xα calculations for $[(SR)_2 Fe_{(1)} S_2 Fe_{(2)} (SR)_2]^{3-}$ together with corresponding values which have been derived from the spin-Hamiltonian analysis of measured Mössbauer spectra (4.2 K) and from EPR measurements of reduced Fe_2S_2 ferredoxins and synthetic analogs. Calculated values taken from [85]

parameter	iron site	R = H 0.275[a]	R = H (0.270)[a]	R = CH$_3$ 0.275[a]	spin-Hamiltonian analysis, EPR
ΔE_Q (mms^{-1})[b]	ferric Fe$_{(1)}$	-1.38	(-1.33)	-1.56	± 0.60 to ± 0.80[c,k]
	ferrous Fe$_{(2)}$	3.09	(2.95)	3.16	± 2.60 to ± 3.20[c,k]
V_{zz}	ferric Fe$_{(2)}$	$\perp Fe_{(1)}$-$Fe_{(2)}$		$\perp Fe_{(1)}$-$Fe_{(2)}$	
	ferrous Fe$_{(2)}$	0.75	(0.38)	0.73	0.6 ± 0.3[d], 0[k]
	ferric Fe$_{(1)}$	0.55	(0.24)	0.59	0.0 ± 0.20[d], 0[k]
η	ferrous Fe$_{(2)}$	-0.43[e]	(-0.48)[e]	-0.41[e]	-0.24 to -0.35[f], -0.41[g], 0.43[k]
$\Delta\delta$ (mms^{-1})					
g'_x[h]		1.87	(1.90)	1.86	1.89[i], 1.94[i], 1.80[k]
g'_y		1.90	(1.93)	1.88	1.96[i], 1.94[i], 1.90[k]
g'_z		1.99	(1.99)	1.98	2.05[i], 2.01[i], 2.02[k]
$A'_x/g_n\beta_n$ (T)[h]	ferric Fe$_{(1)}$	-35.6	-22.3	-35.2	-37 ± 1[i], -41 ± 2[j], -40.5 ± 2.0[k]
	ferrous Fe$_{(2)}$	6.2	-3.0	4.6	8 ± 4[i], 10 ± 2[j], 8 ± 2[k]
$A'_y/g_n\beta_n$ (T)[h]	ferric Fe$_{(1)}$	-35.6	-22.3	-35.2	-35.6 ± 1.8[i], -36.4 ± 1.1[j], -36.4 ± 2.0[k]
	ferrous Fe$_{(2)}$	19.4	(10.9)	17.5	12.2 ± 4[i], 15.2 ± 2.9[j], 10.2 ± 1.1[k]
$A'_y/g_n\beta_n$ (T)[k]	ferric Fe$_{(1)}$	-35.6	-22.3	-35.2	-30.5 ± 1.1[i], -31.2 ± 1.4[j], -31.2 ± 2.9[k]
	ferrous Fe$_{(2)}$	33.9	(21.1)	32.1	25.6 ± 1.4[i], 25.4 ± 1.1[j], 24.0 ± 1.1[k]

[a] $Fe_{(1)}$-$Fe_{(2)}$ distance in nm. The values in parentheses refer to the model system in which only the $Fe_{(2)}$-S distances are increased to 0.236 nm. The model with R = CH$_3$ is based on the real synthetic analog of Mayerle et al. (1975) J. Am. Chem. Soc. 97: 1032.

[b] Sign of ΔE_Q corresponds to sign of V_{zz}.

[c] Taken from Mascharak et al. (1981) J. Am. Chem. Soc. 103: 6110; Anderson et al. (1975) Biochim. Biophys. Acta 408:306; Sands et al. (1975) Quart. Rev. Biophys. 7: 443; Münck et al. (1972) Biochem. 11: 855; Dunham et al. [76]; in some cases the sign of V_{zz} is reported to be positive, in some negative, or it is only marginally preferred over the opposite sign.

[d] Taken from Anderson et al.; Sands et al.; Dunham et al. [76]; a change in sign and orientation of V_{zz} completely changes η.

[e] Calculated from $\Delta\rho(0)$ according to $\Delta\delta = \alpha\Delta\rho(0)$ with a $\alpha = -0.22$ mms^{-1}a^3. (Trautwein et al.) [82].

[f] Taken from Anderson et al.; Sands et al.; Johnson [75]; Münck et al.; Dunham et al. [76]; various ferredoxins.

[g] Taken from Mascharak et al.; synthetic analog.

[h] Calculated values are derived by the vector-coupling model ($\tilde{g}' = (7/3)\tilde{g}_1 - (4/3)\tilde{g}_2$ and $\tilde{A}'_1 = (7/3)\tilde{A}_1$, $\tilde{A}'_2 = -(4/3)\tilde{A}_2$) from individual contributions of Fe$_{(1)}$ and Fe$_{(2)}$, the latter being obtained from the MO electronic structure (Trautwein et al.) [85].

[i] Taken from Dunham et al. [76]; spinach ferredoxin.

[j] Taken from Münck et al., putida ferredoxin.

[k] Taken from Fee et al. [53]; Rieske center in T. thermophilus. A series of synthetic analogs yield similar g-values of $g_{av} \sim 1.91$ (Bertrand et al. [93]).

5.3.3.1 The Ferromagnetic Case (J > 0)

In general, sign and magnitude of the exchange-coupling constant J is significantly dependent on the structural details of the cluster under study [104, 105] (and references therein). Also the simple case presented in Fig. 30 qualitatively illustrates the structure dependence of the crossover from antiferromagnetic to ferromagnetic behaviour. Among the rare binuclear iron clusters which exhibit ferromagnetic behaviour are the recently synthesized molecules [106] with a $Fe_2(OR)_2$-bridge unit containing Fe(III)–Fe(III), Fe(III)–Fe(II) and Fe(II)–Fe(II), though the observed ferromagnetic exchange coupling is very small, i.e. $1.2\ cm^{-1}$, $1.2\ cm^{-1}$, and $8.6\ cm^{-1}$, respectively, as derived from the temperature-dependent magnetic susceptibilities. The ferromagnetic coupling of the valence-trapped Fe(III)–Fe(II) system at low temperature was also demonstrated by EPR and Mössbauer measurements [107].

The X-band EPR spectrum (not shown) exhibits broad resonances centered at $g^{eff} = 6.4$ and 11. From variable-temperature studies it was concluded that these resonances originate from the first ($g^{eff} \sim 11$) and second excited Kramers doublet of an $S = 9/2$ system with $D \cong 1.5\ cm^{-1}$ and $E/D \sim 0.15$–0.33. For this

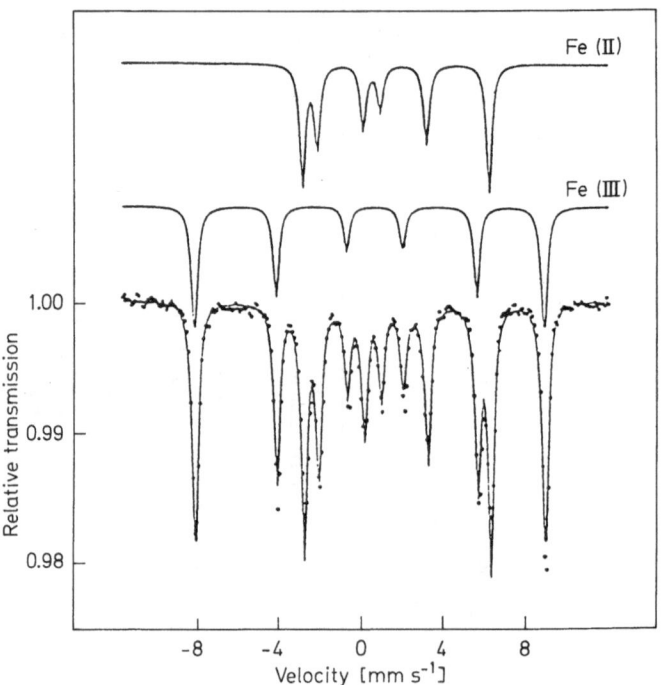

Fig. 33. Experimental Mössbauer spectrum of polycrystalline $(Et_4N)\ [Fe_2(salmp)_2]$ at 1.5 K. Solid lines are simulations using a $S = 9/2$ spin Hamiltonian. Spectral decomposition into contributions from Fe(II) and Fe(III) are shown separately. Taken from [107]

range of E/D, an S = 9/2 system has a ground-state doublet with uniaxial magnetic properties ($g_x^{eff} \sim g_z^{eff} \sim 0$, $g_y^{eff} \approx 17.5$). Such doublets produce Mössbauer spectra that exhibit a six-line pattern for each Fe site. The exceptionally well-resolved spectrum in Fig. 33 is consistent with this expectation. Spectral simulations confirm the S = 9/2 spin state and yield the following information [107]: (i) The two Fe sites occur in an 1:1 occupation ratio. (ii) The quadrupole splitting $\Delta E_Q = 0.92$ mms^{-1}, the isomer shift $\delta(\alpha\text{-Fe}) = 0.55$ mms^{-1}, and the magnetic hyperfine coupling constant A = 21 T are typical of high-spin Fe(III), whereas the parameters of the other site, $\Delta E_Q = 2.35$ mms^{-1}, $\delta(\alpha\text{-Fe}) = 1.12$ mms^{-1}, $A_y = 13.1$ T, are characteristic of high-spin Fe(II), both being hexacoordinated by oxygen and nitrogen ligands. (iii) At 100 K the electronic relaxation is fast, and the Mössbauer spectrum consists of two quadrupole doublets characteristic of localized Fe(III) and Fe(II) sites. At somewhat higher temperatures, a new doublet appears; its intensity increases gradually as the temperature is raised. This doublet exhibits exactly the average of the isomer shift of Fe(III) and Fe(II) sites, which was reported to reflect valence-detrapping (transitions between Fe(III) and Fe(II) that are fast on the Mössbauer time scale, 10^{-7} s, but slow on the vibrational time scale, 10^{-13} s) or actual valence delocalization.

5.3.3.2 Competition Between Antiferromagnetic Heisenberg-Exchange Interaction and Double-Exchange Interaction

In mixed-valence complexes like Fe(III)–Fe(II) dimers it might occur that the "excess" electron (which is added to reduce the Fe(III)–Fe(III) dimer) is not trapped at one of the two paramagnetic sites but delocalized between these sites (class III delocalized-valence compounds [108]). This situation leads to a new type of electrostatic interaction mechanism in addition to the superexchange interaction, viz. double-exchange interaction as described by Eq. (48). Assuming that the structure of the investigated system predicts J < 0 (antiferromagnetic spin coupling), the actual system spin of the ground state is dependent on the ratio J/|B|. For J/|B| > −2/9 the ground-state spin S_t is 9/2, whereas for J/|B| < −2/9 ground-state spins are found in the order $S_t = 7/2, 5/2, \ldots, 1/2$, with decreasing J/|B| (Fig. 34). The recently synthesized compound [L$_2$Fe$_2$(μ-OH)$_3$](ClO$_4$)$_2 \cdot$ 2CH$_3$OH $\cdot$ 2H$_2$O   (L = N,N',N''-trimethyl-1,4,7,-triazacyclononane) is an example which falls in the category J/|B| > −2/9 with S_t = 9/2 [109, 110].

The Mössbauer spectrum of [L$_2$Fe$_2$(μ-OH)$_3$]$^{2+}$ [110] (all measurements were performed on powder) at 4.2 K in zero applied field exhibits a single quadrupole doublet, which indicates that the two iron sites of the dimer are indistinguishable, at least within the Mössbauer time scale of about 10^{-7} s. The isomer shift $\delta(\alpha\text{-Fe}) = 0.74$ mms^{-1} lies between the values $\delta = 0.35$–0.60 mms^{-1} and $\delta = 0.90$–1.30 mms^{-1} observed, respectively, in ferric and ferrous high-spin iron, coordinated by six oxygen or nitrogen ligands

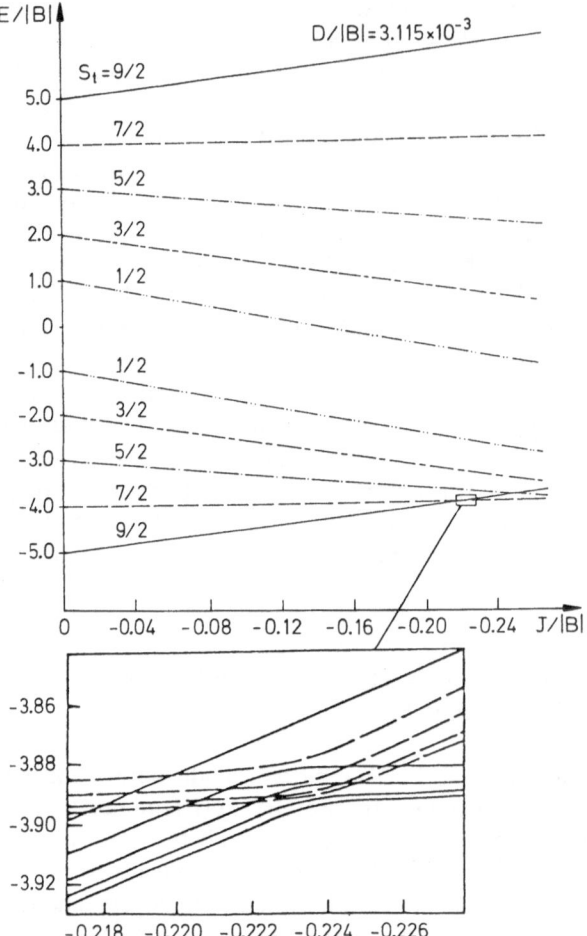

Fig. 34. Energies of spin states of the delocalized-valence $Fe^{2.5+} - Fe^{2.5+}$ dimer as a function of $J/|B|$ calculated with the local spin Hamiltonian $\mathscr{H} = D_1[S_{1z}^2 - 35/12 + E_1(S_{1x}^2 - S_{1y}^2)/D_1] + D_2[S_{2z}^2 - 2 + E_2(S_{2x}^2 - S_{2y}^2)/D_2] - J\vec{S}_1 \cdot \vec{S}_2 + T$ and using the parameters $D = D_1 = D_2 = 3.115 \times 10^{-3}|B|$, $E = E_1 = E_2 = 0$. Taken from [110]

[111, 41]. This observation indicates that the dimer is a delocalized-valence pair of the form $Fe_A^{2.5+}$–$Fe_B^{2.5+}$, in which one electron is equally distributed over the two iron sites.

For the determination of the hyperfine-interaction parameters, a series of Mössbauer spectra, in different applied fields at varying temperatures, were recorded and analyzed with a spin Hamiltonian based on $S_t = 9/2$ [10]. The low-temperature spectra are presented in Fig. 35 together with the corresponding simulations, using the parameters in Table 3. Confirmation for the ground-state spin $S_t = 9/2$ is provided by the magnitude of the efffective moment

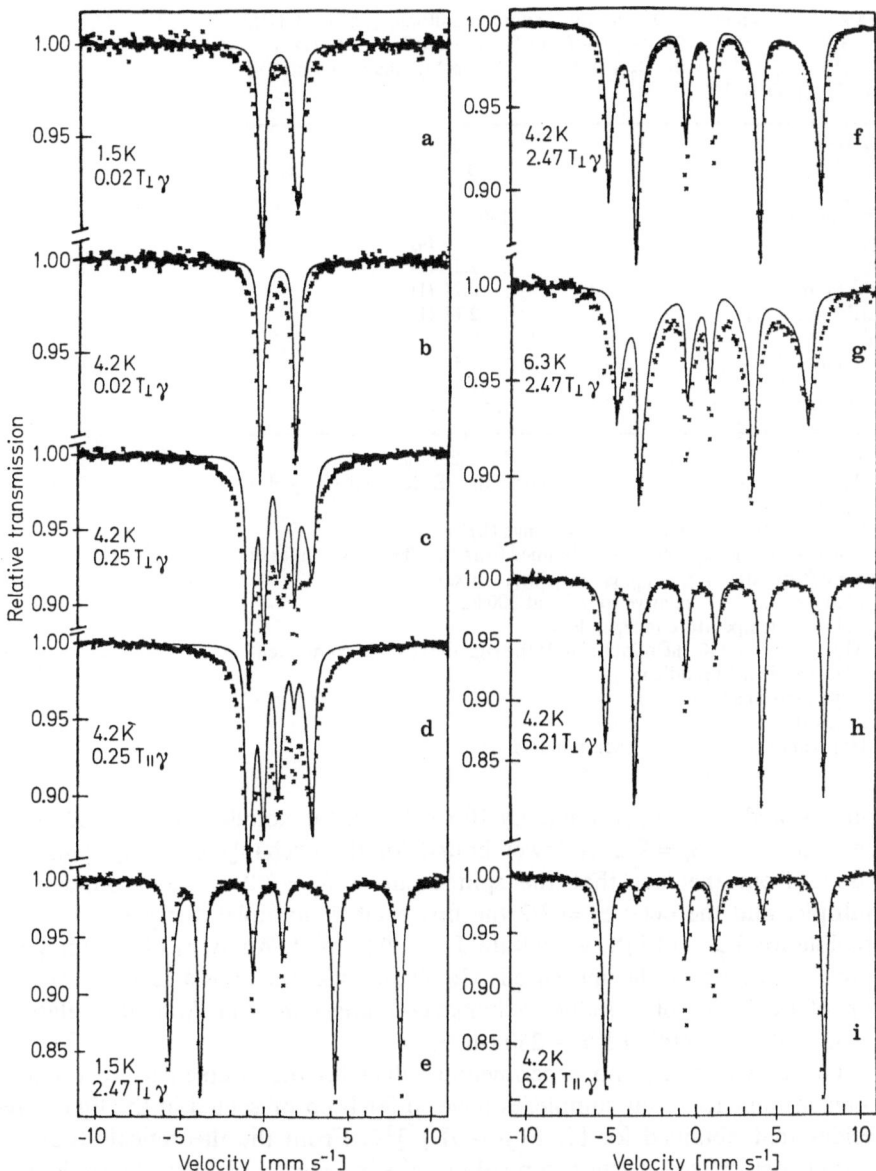

Fig. 35a–i. Experimental Mössbauer spectra of $[L_2Fe_2(\mu\text{-}OH)_3]^{2+}$ recorded at temperatures and in applied fields as indicated. *Solid lines* are simulations using local-spin Hamiltonian and parameters given in Table 3

$\mu^{eff} \approx 10\ \beta$, deduced from magnetic susceptibility measurements, and from the effective g-factor $g_\perp^{eff} \approx 10$, obtained from an EPR spectrum at 2.5 K [110].

The value $B = 1300\ cm^{-1}$ was obtained by identifying the intense absorption band at $\lambda = 758$ nm [109] in the UV-VIS spectrum with the Laporte and

Table 3. Parameters used in local-spin Hamiltonian $\mathscr{H} = D_1[S_{1z}^2$
$- 35/12 + E_1(S_{1x}^2 - S_{1y}^2)/D_1] + D_2[S_{2z}^2 - 2 + E_2(S_{2x}^2 - S_{2y}^2)/D_2]$
$+ \beta\vec{B}\cdot\vec{g}\cdot(\vec{S}_1 + \vec{S}_2) - J\vec{S}_1\cdot\vec{S}_2 + T$ for simulating Mössbauer spectra of
$[L_2Fe_2(\mu\text{-OH})_3]^{2+}$

D^a (cm^{-1})	4.0	(5)
$E/D^{a,b}$	0	
J^c (cm^{-1})	> -250	
B^c (cm^{-1})	1300	
$g_{x,y}{}^d$	2.04	(4)
$g_z{}^d$	2.3	(1)
$\delta^{e,f}$ (mms^{-1})	0.74	(1)
$\Delta E_Q^{f,g}$ (mms^{-1})	-2.14	(1)
η^h	0	
Γ^i (mms^{-1})	0.3	
$A_{x,y}/g_n\beta_n^j$(T)	-21.2	(5)
$A_z/g_n\beta_n^j$(T)	-27	(2)

[a] $D_1 = D_2 = D = 2.22D_{9/2}$; $E_1 = E_2 = E = 2.22E_{9/2}$.
[b] The value zero was taken because the EPR analysis yields
$E_{9/2}/D_{9/2} < 0.01$.
[c] Superexchange (J) and double-exchange (B) parameters.
[d] The values of g_x, g_y, and g_z are obtained from the effective g′ values of
the EPR spectra by $g_z = g'_{||}$, $g_i = (1/5)g'_\perp$, i = x, y.
[e] Measured at 4.2 K relative to α-Fe at 300 K.
[f] Taken as temperature independent.
[g] Measured at 4.2 K; the minus sign is the sign of the main component
of the electric field gradient.
[h] Asymmetry parameter.
[i] Linewidth.
[j] Hyperfine coupling tensor components.

spin-allowed transitions of energy 10B between the gerade and ungerade system-spin states $S_t = 9/2$. A lower bound for the exchange-coupling constant
follows from the fact that the spin decuplet $S_t = 9/2$ is the ground-state
multiplet and the octet $S_t = 7/2$ the first excited multiplet. Since this is only
possible for $J > -(2/9)B$, we obtain $J > -288$ cm^{-1}. Analyses of the temperature variations of the measured Mössbauer spectra allows further restric-
tion of the range of possible exchange-coupling constants from the relation
$J = (2/9)(\Delta\text{-B})$, yielding $J > -250$ cm^{-1}.

To our knowledge, no experimental values for the double-exchange parameter in binuclear iron complexes have so far been presented in the literature
besides that obtained for $[L_2Fe_2(\mu\text{-OH})_3]^{2+}$. From the theoretical study of
partially delocalized-valence iron-sulfur units in reduced plant-type ferredoxins
[84] a double-exchange parameter of 500 cm^{-1} was deduced, which is of the
same order of magnitude as the experimental value for B in $[L_2Fe_2(\mu\text{-OH})_3]^{2+}$
The values B = 500 cm^{-1} calculated for distance Fe–Fe = 0.27 nm in reduced
ferredoxin and B = 1300 cm^{-1} measured for Fe–Fe = 0.25 nm in $[L_2Fe_2(\mu$-
OH)$_3]^{2+}$ suggest stronger double-exchange interaction for shorter iron–iron
separations. It should be noted, however, that the double-exchange parameter
may also depend on the nature of ligand bridges between the iron sites, i.e. sulfur
in ferredoxin and hydroxo-oxygen in $[L_2Fe_2(\mu\text{-OH})_3]^{2+}$.

The overall conclusion of this discussion is that in mixed-valence dimers, even under circumstances which favour antiferromagnetic exchange coupling ($J < 0$), the spins of the two paramagnetic sites may tend to align parallel; this is the case when double exchange dominates superexchange as illustrated in Fig. 34 for $J/|B| > -2/9$.

5.4 Iron Clusters with Three and Four Metal Sites

5.4.1 Three- and Four-Iron Clusters Containing a Delocalized Mixed-Valence Dimer

Delocalized-mixed valence dimers containing an isolated Fe_2S_2 unit have not been observed thus far. The "synthetic analog" discussed in Sect. 5.3.3.2 is a μ-hydroxo-bridged two-iron cluster, which is believed to exhibit the same gross spectroscopic features as the corresponding hypothetical sulfide-bridged cluster. There is strong support for the conclusion, that in iron-sulfur proteins the reduced Fe_3S_4 cluster as well as the Fe_4S_4 cubanes (in all three accessible oxidation states) contain a delocalized mixed-valence dimer Fe_2S_2 [112–114]. This conclusion is partly based on the interpretation of measured isomer shifts and quadrupole splittings (Table 4). The delocalized-mixed valence dimer ($Fe^{2.5+}$–$Fe^{2.5+}$) exhibits isomer shifts just in between the values characteristic for $Fe^{3+}S_4$ (~ 0.3 mms^{-1}) and $Fe^{2+}S_4$ (~ 0.6 mms^{-1}).

The three Fe sites of the Fe_3S_4 core in ferredoxin II have been studied under conditions where the fourth site is occupied by Fe^{2+}, Co^{2+} or Zn^{2+} [115–118] (Fig. 36). Thus, one may ask to what extent the properties of the delocalized pair and the single (third) site are modified by the occupation of the fourth site. Using the isomer shift again as oxidation-state indicator it is seen from Table 4 that the delocalized pair is essentially unaltered in the series $[Fe_3S_4]^0$, $[Fe_4S_4]^{2+}$ $[CoFe_3S_4]^{2+}$. On the other hand, the additional ferric site becomes more ferrous through the series. In fact, in $[Fe_4S_4]^{2+}$ it belongs to a second de-localized pair, the two sites of which, however, can be discerned by their distinct

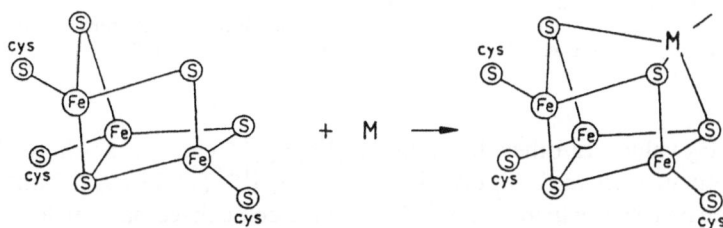

Fig. 36. Putative structure of Fe_3S_4 cluster and illustration of conversion into MFe_3S_4 by substituting M = Fe, Co, Zn. EXAFS data of the Fe_3S_4 cluster suggest a compact core with an Fe–Fe distance of 0.27 nm (Antonio et al. (1982) J. Biol. Chem. 257: 6646), just as observed for Fe_4S_4 cubanes

Table 4. Isomer shifts $\delta(\alpha\text{-Fe})$ and quadrupole splittings ΔE_Q of three-iron and four-iron clusters in various oxidation states

cluster	valence		δ (mms^{-1})	ΔE_Q (mms^{-1})	Fe sites	cluster spin
$[Fe_3S_4]^0$	$Fe^{2.5+}$	$- Fe^{2.5+}$	0.46	1.47	2	$\left.\begin{array}{c} \\ \\ \end{array}\right\}$ $S = 2$
a	Fe^{3+}		0.32	0.52	1	
$[Fe_4S_4]^{1+}$	$Fe^{2.5+}$	$- Fe^{2.5+}$	0.50	1.18	2	$\left.\begin{array}{c} \\ \\ \end{array}\right\}$ $S = 1/2$
b	Fe^{2+}	$- Fe^{2+}$	0.60	1.82	2	
$[Fe_4S_4]^{2+}$	$Fe^{2.5+}$	$- Fe^{2.5+}$	0.42	1.08	2	$\left.\begin{array}{c} \\ \\ \end{array}\right\}$ $S = 0$
c	$Fe^{2.5+}$	$- Fe^{2.5+}$	0.42	1.08	2	
$[Fe_4S_4]^{2+}$	$Fe^{2.5+}$	$- Fe^{2.5+}$	0.43	1.14	2	$\left.\begin{array}{c} \\ \\ \end{array}\right\}$ $S = 0$
d	$Fe^{2.5+}$	$- Fe^{2.5+}$	0.43	1.14	2	
$[Fe_4S_4]^{3+}$	$Fe^{2.5+}$	$- Fe^{2.5+}$	0.40	1.03	2	$\left.\begin{array}{c} \\ \\ \end{array}\right\}$ $S = 1/2$
e	Fe^{3+}	$- Fe^{3+}$	0.29	0.88	2	
$[Fe_4S_4]^{1+}$	$Fe^{2.5+}$	$- Fe^{2.5+}$	0.51	1.07	2	$\left.\begin{array}{c} \\ \\ \end{array}\right\}$ $S = 1/2$
f	Fe^{2+}	$- Fe^{2+}$	0.60	1.67	2	
$[Fe_4S_4]^{2+}$	$Fe^{2.5+}$	$- Fe^{2.5+}$	0.45	1.32	2	$\left.\begin{array}{c} \\ \\ \\ \end{array}\right\}$ $S = 0$
f	$Fe^{2.5+}$	$- Fe^{2.5+}$	0.45	1.32	1	
			0.41	0.55	1	
$[ZnFe_3S_4]^{1+}$	$Fe^{2.5+}$	$- Fe^{2.5+}$	0.52	1.5	1	$\left.\begin{array}{c} \\ \\ \\ \end{array}\right\}$ $S = 5/2$
				1.7	1	
f	Fe^{2+}	$- Zn^{2+}$	0.62	2.7	1	
$[CoFe_3S_4]^{1+}$	$Fe^{2.5+}$	$- Fe^{2.5+}$	0.53	1.28	2	$\left.\begin{array}{c} \\ \\ \end{array}\right\}$ $S > 0$
f	Fe^{2+}	$- Co^{2+}$	0.53	1.28	1	
$[CoFe_3S_4]^{2+}$	$Fe^{2.5+}$	$- Fe^{2.5+}$	0.44	1.35	2	$\left.\begin{array}{c} \\ \\ \end{array}\right\}$ $S = 1/2$
f	Fe^{3+}	$- Co^{2+}$	0.35	1.1	1	

[a] Ferredoxin II from *Desulfovibrio gigas*. Taken from [112].
[b] Reduced ferredoxin from *B. stearothermophilus*. Taken from [113].
[c] Oxidized ferredoxin from *B. stearothermophilus*. Taken from [113].
[d] Reduced *Chromatium* HiPIP. Taken from [114].
[e] Oxidized *Chromatium* HiPIP. Taken from [114].
[f] $[MFe_3S_4]^{1+,2+}$ clusters obtained by incubating ferredoxin II from *Desulfovibrio gigas* in the presence of excess $M = Fe^{2+}, Co^{2+}, Zn^{2+}$. Taken from [115–118].

ΔE_Q values. In other diamagnetic $[Fe_4S_4]^{2+}$ clusters, as shown in Table 4, the four sites are indistinguishable. In $[Fe_4S_4]^{1+}$, $[CoFe_3S_4]^{1+}$ and $[ZnFe_3S_4]^{1+}$ the iron sites which do not belong to the delocalized pair are formally Fe^{2+}. The values in Table 4 indicate that isomer shift and quadrupole splitting of the delocalized pair may vary from one compound to the other in the range 0.40–0.53 mms^{-1} and 1–1.7 mms^{-1}, respectively.

We now turn to the discussion of magnetic properties of some of these clusters.

5.4.1.1 $[Fe_3S_4]^\circ$

The reduced Fe_3S_4 cluster of ferredoxin II has cluster spin S = 2 as derived from MCD measurements [119]. This suggests that information about the zero-field splitting and magnetic hyperfine interactions can be obtained by studying the material in strong applied fields. Therefore spectra in fields up to 6 T over a wide range of temperatures have been recorded [112, 118]. A typical spectrum, taken in a field of 1 T at 1.3 K, is shown in Fig. 37 together with a zero-field spectrum recordéd at 4.2 K. The effective magnetic hyperfine coupling parameters (Table 5), obtained from a spin-Hamiltonian analysis of these spectra, are used to outline the magnetic properties of the $[Fe_3S_4]^0$ cluster. The two irons of the $Fe^{2.5+}$–$Fe^{2.5+}$ pair are indistinguishable, comparable to the situation discussed in section 5.3, hence, the same tensor $(\tilde{A}'_{12}/g_n\beta_n)$ is obtained for both sites. The tensor of the Fe^{3+} site $(\tilde{A}'_3/g_n\beta_n)$ has *positive* components. This observation is highly significant. Since a tetrahedral sulfur environment always yields a high-spin configuration, the A-tensor $(\tilde{A}/g_n\beta_n)$ of a ferric FeS_4 site is dominated by

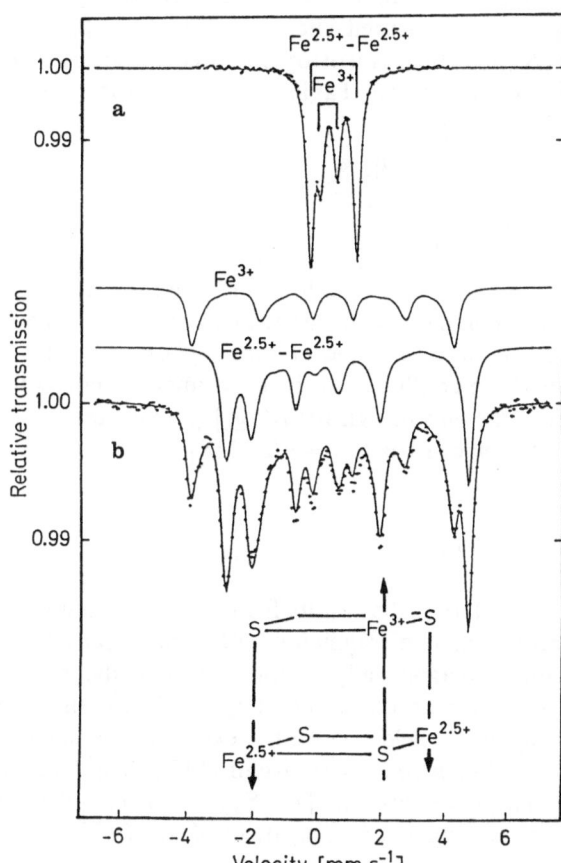

Fig. 37. Mössbauer spectra of the reduced Fe_3S_4 cluster of ferredoxin II recorded at 4.2 K in zero field (**a**) and at 1.3 K in parallel applied field of 1 T (**b**). The solid lines in (**b**) are theoretical curves for the delocalized pair and the Fe^{3+} site. The sum is drawn through the experimental data. Taken from [118]. The insert gives a scheme of the proposed spin-coupling: $S = S_{12} + S_3$ $= 9/2 - 5/2 = 2$

Table 5. Effective magnetic hyperfine coupling tensors[a] $\tilde{A}'_{12}/g_n\beta_n$ and $\tilde{A}'_3/g_n\beta_n$ of reduced ferredoxin II [118]

site	$\dfrac{A'_x(T)}{g_n\beta_n}$	$\dfrac{A'_y(T)}{g_n\beta_n}$	$\dfrac{A'_z(T)}{g_n\beta_n}$	$\dfrac{A'_{av}(T)}{g_n\beta_n}$
$Fe^{2.5+} - Fe^{2.5+}$	-14.9	-14.9	-11.6	-13.8
Fe^{3+}	9.9	11.6	12.6	11.3

[a] Effective tensors $\tilde{A}'$ are related to local tensors $\tilde{A}$ by $\tilde{A}'\langle\vec{S}_t\rangle/g_n\beta_n = \tilde{A}\langle\vec{S}_{local}\rangle/g_n\beta_n$

the Fermi contact term, which is isotropic and *negative*. This implies that the local spin of the Fe^{3+} site ($S_3 = 5/2$) is aligned to the system spin $S = 2$ in such a way, that its projection onto the direction of $\vec{S}_t$ is negative. In a Mössbauer experiment one measures $\tilde{A}'_3\langle\vec{S}_t\rangle/g_n\beta_n$, where $\langle\vec{S}_t\rangle$ is the expectation value of the system spin $\vec{S}_t$ and $A'_3/g_n\beta_n$ the observed magnetic hyperfine coupling constant (which for simplicity is taken as the average of $A'_x/g_n\beta_n$, $A'_y/g_n\beta_n$ and $A'_z/g_n\beta_n$ (Table 5). From the equivalence $\tilde{A}'_3\langle\vec{S}_t\rangle/g_n\beta_n = \tilde{A}_3\langle\vec{S}_3\rangle/g_n\beta_n$ one can relate the measured effective value $A'_3/g_n\beta_n$ to the "uncoupled" (local) value $A_3/g_n\beta_n$, which is isotropic and corresponds to $A_3/g_n\beta_n \approx -15$ T for a ferric site (compare with Table 2). From vector coupling, corresponding to Eq. (52), follows

$$\langle\vec{S}_3\rangle = \frac{\vec{S}_3\cdot\vec{S}_t}{S_t^2}\langle\vec{S}_t\rangle, \tag{54}$$

and therefore $\langle\vec{S}_3\rangle = -(5/6)\langle\vec{S}\rangle$ for $S_{12} = 9/2$ and $\langle\vec{S}_3\rangle = -(1/12)\langle\vec{S}\rangle$ for $S_{12} = 7/2$. For $S_{12} = 9/2$ one obtains $A'_3/g_n\beta_n = -(5/6)A_3/g_n\beta_n = +12.5$ T, which compares well with the experimental value $A'_3(av)/g_n\beta_n = +11.3$ T (Table 5). For $S_{12} = 7/2$ one obtains $A'_3/g_n\beta_n = -(1/12)A_3/g_n\beta_n = +1.2$ T, which is clearly in conflict with the experimental data. All other possible values for S_{12} yield the wrong sign for $A'_3/g_n\beta_n$. Thus, the conclusion is that the delocalized dimer indeed has $S_{12} = 9/2$ [118].

5.4.1.2 $[Fe_4S_4]^{2+}$

Extrapolating the results from $[Fe_3S_4]^0$ suggests that the core of $[Fe_4S_4]^{2+}$ consists of two fragments with dimer spins $S_{12} = S_{34} = 9/2$, which are then antiferromagnetically coupled to give the observed $S = 0$ system spin. The Mössbauer spectrum of $[Fe_4S_4]^{2+}$ proteins and synthetics analogs, when applying a field at 4.2 K, exhibits the same magnetic hyperfine pattern [113, 114] as the $S = 0$ system of $[Fe_2S_2]^{2+}$ ferredoxins, as shown in Fig. 27b, which proves that the $[Fe_4S_4]^{2+}$ cluster indeed is a diamagnetic system. The antiferromagnetic arrangement of two $Fe^{2.5+}-Fe^{2.5+}$ dimers in $[Fe_4S_4]^{2+}$ has also been suggested from Xα valence-bond calculations [120] which yield for the coupling between the two $S = 9/2$ dimers the exchange-coupling constant $J \sim -380 \text{ cm}^{-1}$, compared with $J \sim -460 \text{ cm}^{-1}$, as obtained from magnetic susceptibility measurements for the analogs $[Fe_4S_4(SCH_2Ph)_4]^{2-}$ and $[Fe_4S_4(SPh)_4]^{2-}$ [121].

5.4.1.3 $[Fe_4S_4]^{1+}$

On reduction, the $[Fe_4S_4]^{2+}$ cluster gains one electron, which reduces one of its two $Fe^{2.5+}-Fe^{2.5+}$ dimers. Hence, in this picture a $[Fe_4S_4]^{1+}$ cluster consists of one remaining $Fe_{(1)}^{2.5+}-Fe_{(2)}^{2.5+}$ dimer and of one $Fe_{(3)}^{2+}-Fe_{(4)}^{2+}$ dimer, with dimer spins $S_{12} = 9/2$ and $S_{34} = 4$, respectively. Since these dimer spins are coupled antiparallel to total spin $S_t = 1/2$, application of a field yields the orientation of spins and magnetic moments as depicted in Fig. 38. With this spin arrangement the relation between g- and A-tensors of the individual iron atoms $Fe_{(i)}$ ($i = 1 \ldots 4$) and their effective counterparts $\tilde{g}'$ and $\tilde{A}'_i$, which are measured by EPR and Mössbauer spectroscopy, becomes more transparent.

Before we continue with this explanation, we will show experimental Mössbauer spectra in Fig. 39 [114] and an EPR spectrum in Fig. 40a [122] of reduced B. stearothermophilus ferredoxin and in Table 6 the corresponding spin-Hamiltonian parameters, which have been obtained by simulating the Mössbauer spectra in the framework of total (system) spin $S_t = 1/2$, i.e. assuming strong antiferromagnetic exchange interaction between the two dimer spins $S_{12} = 9/2$ and $S_{34} = 4$. Since positive $A'/g_n\beta_n$-values correspond to positive hyperfine fields ($B^{hf} = -A'\langle S_t\rangle/g_n\beta_n$) at the ^{57}Fe nucleus and vice versa, the observation of positive and negative $A'/g_n\beta_n$-values confirms the existence of two opposing fields within the $[Fe_4S_4]^{1+}$ center and, hence, antiparallel spin coupling. Consistent with this picture is the field dependence of the dimer-related subspectra in Fig. 39. Inspection of the isomer shifts and the anisotropy behavior of $\tilde{A}'/g_n\beta_n$ in Table 6 identifies the positive $\tilde{A}'/g_n\beta_n$-components as those corresponding to the $Fe^{2+}-Fe^{2+}$ pair.

We now turn back to the spin-coupling model for the $S_t = 1/2$ system, depicted in Fig. 38 and apply an analysis of g'- and A'-tensors, which is an

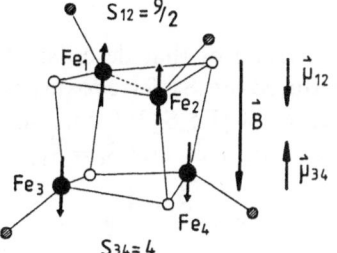

Fig. 38. Model for the spin arrangement in $[Fe_4S_4]^{1+}$ in an applied field $\vec{B}$. The *dashed line* indicates the delocalized electron within the $Fe^{2.5+}-Fe^{2.5+}$ dimer. Total spin is $S_t = 1/2$

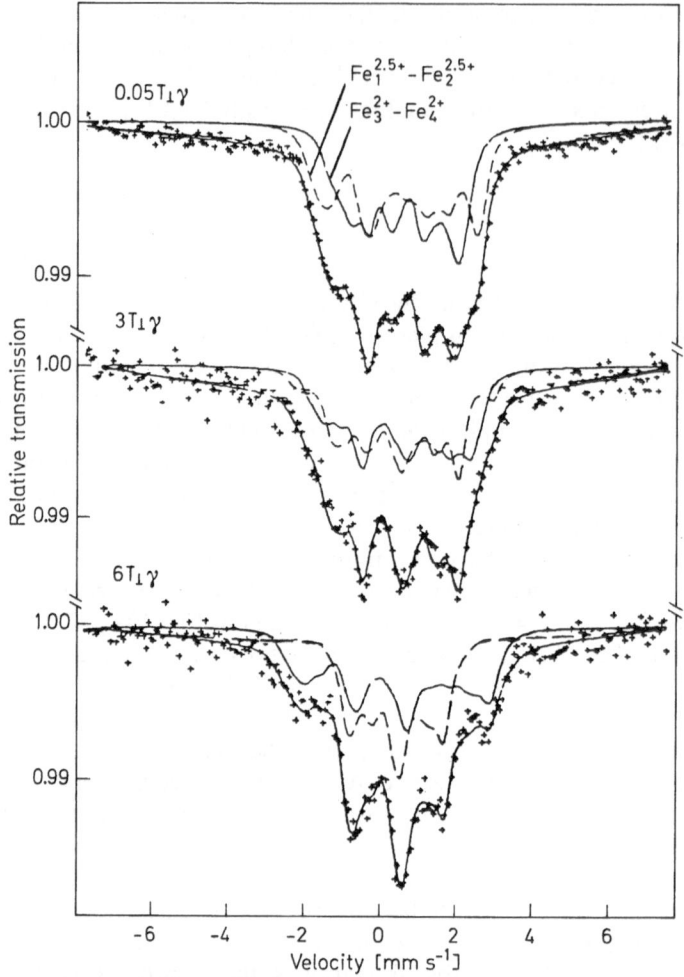

Fig. 39. Experimental Mössbauer spectra and their spin-Hamiltonian simulation (parameters in Table 6) of reduced *B. stearothermophilus* ferredoxin at 4.2 K and externally applied fields as indicated. The overall magnetic splitting (nuclear Zeeman interaction) of each subspectrum is proportional to the effective field sensed by the ^{57}Fe nucleus: $\vec{B}^{eff} = \vec{B}^{hf} + \vec{B}^{ext}$. Since $\vec{B}^{hf}_{12} < 0$ and $B^{hf}_{34} > 0$, the overall magnetic splitting for the $Fe^{2.5+}$–$Fe^{2.5+}$ pair decreases and that for the Fe^{2+}–Fe^{2+} pair increases with increasing B^{ext}. Taken from [114]

extension of that described for the simpler case of the two coupled spins in reduced Fe_2S_2 ferredoxins (section 5.3.1). According to Fig. 38 the $S = 1/2$ center of $[Fe_4S_2]^{1+}$ consists of two spin-coupled dimers with $\vec{S}_{12} = \vec{S}_1 + \vec{S}_2$, $\vec{S}_{34} = \vec{S}_3 + \vec{S}_4$ and $\vec{S}_t = \vec{S}_{12} + \vec{S}_{34}$. Spin-coupling yields

$$\tilde{g}' = \frac{\vec{S}_{12} \cdot \vec{S}_t}{S_t^2} \tilde{g}_{12} + \frac{\vec{S}_{34} \cdot \vec{S}_t}{S_t^2} \tilde{g}_{34}, \tag{55}$$

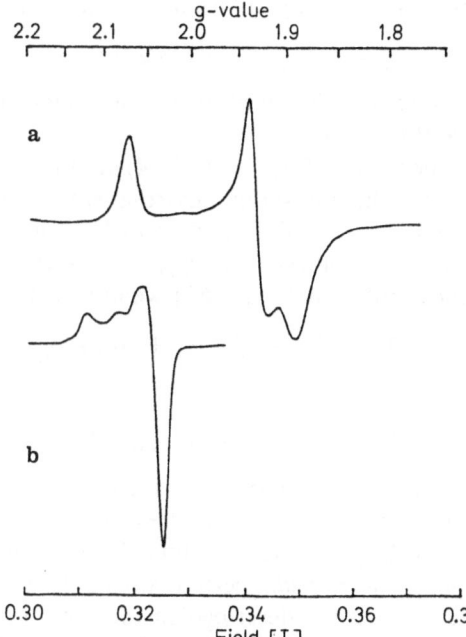

Fig. 40. Experimental EPR spectra of Fe_4S_4 proteins, (a) reduced B. *stearothermophilus* ferredoxin at 12 K, (b) oxidized *Chromatium* HiPIP at 20 K. Taken from [122]

Table 6. Spin-Hamiltonian parameters obtained from simulating Mössbauer spectra at 4.2 K of reduced B. *stearothermophilus* ferredoxin (Fig. 39). Taken from [114]

Component	$\delta(\alpha\text{-Fe})$ (mms^{-1})	ΔE_Q (mms^{-1})	η	$\dfrac{A'_x(T)}{g_n\beta_n}$	$\dfrac{A'_y(T)}{g_n\beta_n}$	$\dfrac{A'_z(T)}{g_n\beta_n}$
$Fe_{(1)}^{2.5+} - Fe_{(2)}^{2.5+}$	0.50	1.32	0.78	-23.0	-23.6	-20.0
$Fe_{(3)}^{2+} - Fe_{(4)}^{2+}$	0.58	1.89	0.32	$+19.2$	$+9.8$	$+6.3$

with $\tilde{g}_{12(34)}$ derived from adequate application of Eq. (51). $\vec{S}_1$ and $\vec{S}_2$ are site spins for $Fe^{3+}-Fe^{2+}$, and $\vec{S}_3$ and $\vec{S}_4$ are site spins for $Fe^{2+}-Fe^{2+}$, so $S_1 = 5/2$, $S_2 = 2$, and $S_3 = S_4 = 2$. (The first two spin values can be interchanged without affecting the arguments.) With these values and with $S_{12} = 9/2$, $S_{34} = 4$, $S_t = 1/2$, and additionally making use of eq. (52), one obtains

$$\tilde{g}' = (55/27)\tilde{g}_1 + (44/27)\tilde{g}_2 - (4/3)\tilde{g}_3 - (4/3)\tilde{g}_4, \tag{56}$$

and accordingly

$$\tilde{A}'_1 = (55/27)\tilde{A}_1, \quad \tilde{A}'_2 = (44/27)\tilde{A}_2, \quad \tilde{A}'_{3(4)} = -(4/3)\tilde{A}_{3(4)}. \tag{57}$$

Ferric high-spin iron has g-values which are isotropic and equal to 2, while ferrous high-spin iron has g-values which are anisotropic with $g_z = 2$ and $g_{x,y} = 2 + \Delta$ (assuming axial symmetry) as described in section 5.3.1. Putting these single-atom g-values in Eq. (56) yields

$$g'_z = 2, \; g'_{x,y} = 2 - (28/27)\Delta. \tag{58}$$

This gives g'-values with an average of less than 2 as observed experimentally by EPR (Fig. 40a).

The values of $\tilde{A}_i$ ($i = 1 \ldots 4$) are those for individual ferric and ferrous iron in a tetrahedral sulfur environment. Using the isotropic value $\tilde{A}_1/g_n\beta_n = (-14.5\,T, \, -14.5\,T, \, -14.5\,T)$ for Fe^{3+} in oxidized Fe_3S_4 clusters [118] and the anisotropic values $A_{2,3,4}/g_n\beta_n = (-16.8\,T, \, -11.5\,T, \, -4.7\,T)$ for Fe^{2+} in reduced rubredoxin [56, 62] we arrive at average values

$$\tilde{A}'_{12}/g_n\beta_n \sim (1/2)(\tilde{A}'_1 + \tilde{A}'_2)/g_n\beta_n = (55/54)\tilde{A}_1/g_n\beta_n + (44/54)\tilde{A}_2/g_n\beta_n$$

$$= (-28.5\,T, \, -24.7\,T, \, -18.6\,T),$$

$$\tilde{A}'_{34}/g_n\beta_n = -(4/3)A_{3(4)}/g_n\beta_n = (22.4\,T, \, 15.3\,T, \, 6.8\,T),$$

which are in reasonable agreement with the values given in Table 6.

After having completed the discussion of the relatively simple situation, that the $[Fe_4S_4]^{1+}$ center exhibits cluster spin $S_t = 1/2$ as in the case of reduced *B. stearothermophilus* ferredoxin [114], we want to emphasize that there are several examples described in the literature which involve $[Fe_4S_4]^{1+}$ centers with "nonstandard" spin $S > 1/2$. Such behaviour has been identified by the combined EPR and Mössbauer investigation in the $2[Fe_4Se_4]^{1+}$ ferredoxin from *C. pasteurianum* [123, 124] (which exhibits even a mixture of cluster spins $S_t = 1/2$, $S_t = 3/2$ and $S_t = 7/2$), in the Fe protein from *A. vinelandii* nitrogenase [125], in amidotransferase from *B. subtilis* [126], and in a series of synthetic analogs such as $[Fe_4S_4(SPh)_4]^{3-}$, $[Fe_4S_4(SPh)_4]^{3-}$ [127] and $[Fe_4Se_4(SC_6H_{11})_4]^{3-}$ [128].

5.4.1.4 $[Fe_4S_4]^{3+}$

On oxidation, the $[Fe_4S_4]^{2+}$ cluster loses one electron. According to the discussion of $[Fe_4S_4]^{2+}$ and $[Fe_4S_4]^{+}$ it is tempting to assume that this extra electron has been given up by one of the $Fe^{2.5+}-Fe^{2.5+}$ pairs which becomes a $Fe^{3+}-Fe^{3+}$ pair, while the other $Fe^{2.5+}-Fe^{2.5+}$ pair remains. This situation is fully reflected by the isomer shifts derived from the Mössbauer spectrum of the oxidized high-potential iron protein (HiPIP) from *Chromatium* [129], which was recorded at 4.2 K in a field of 10 T applied parallel to the γ-beam (Fig. 41). Continuing the analogy between $[Fe_4S_4]^{1+}$ and $[Fe_4S_4]^{3+}$ one would expect strong antiferromagnetic exchange coupling between the two pairs $Fe^{3+}_{(1)}-Fe^{3+}_{(2)}$ ($S_{12} = 5$) and $Fe^{2.5+}_{(3)}-Fe^{2.5+}_{(4)}$ ($S_{34} = 9/2$). Applying the spin-coupling model, discussed for $[Fe_4S_4]^{1+}$, to the proposed situation for $[Fe_4S_4]^{3+}$ ($S_1 = S_2 = S_3 = 5/2$, $S_4 = 2$, $S_{12} = 5$, $S_{34} = 9/2$, $S_t = 1/2$) would yield for the $Fe^{2+}_{(1)}-Fe^{3+}_{(2)}$ pair approximately $A'_{12}/g_n\beta_n \sim (-30\,T, \, -30\,T, \, -30\,T)$ and for the $Fe^{2.5+}_{(2)}-Fe^{2.5+}_{(4)}$ pair $A'_{34}/g_n\beta_n \sim (+23\,T, \, +20\,T, \, +15\,T)$, and additionally

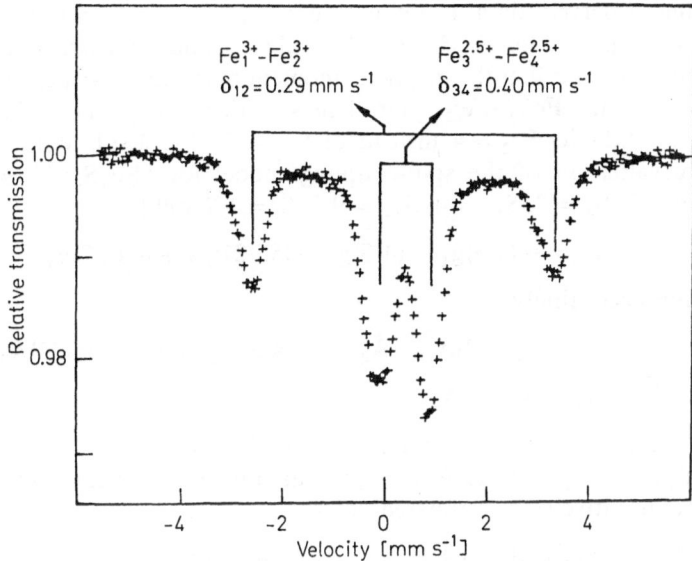

Fig. 41. Experimental Mössbauer spectrum of oxidized *Chromatium* HiPIP at 4.2 K in an applied field of 10 T parallel to the γ-beam. Taken from [129]

Table 7. Spin-Hamiltonian parameters obtained from simulating Mössbauer spectra at 4.2 K of oxidized *Chromatium* HiPIP. Taken from [129]

Component	$\delta(\alpha\text{-Fe})$ (mms^{-1})	ΔE_Q (mms^{-1})	η	$\dfrac{A'_x(T)}{g_n\beta_n}$	$\dfrac{A'_y(T)}{g_n\beta_n}$	$\dfrac{A'_z(T)}{g_n\beta_n}$
$Fe^{3+}_{(1)} - Fe^{3+}_{(2)}$	0.29	0.88	0.4	+13.9	+16.3	+14.0
$Fe^{2.5+}_{(3)} - Fe^{2.5+}_{(4)}$	0.40	1.03	0.9	−20.5	−22.2	−23.6

a set of g'-values with an average of less than 2. These results, however, are in complete contradiction to the data observed for oxidized HiPIP from *Chromatium*, i.e. from the spin-Hamiltonian analysis of field-dependent Mössbauer spectra recorded at 4.2 K and from an EPR spectrum recorded at 20 K (Fig. 40b). The spin-Hamiltonian analysis (Table 7) yields positive(!) sign for $A'_{12}/g_n\beta_n$ and negative(!) sign for $A'_{34}/g_n\beta_n$, and the EPR spectrum exhibits $g_{av} > 2$.

In general a high-spin ferric atom ($S = 5/2$) will have a larger magnetic moment than a high-spin ferrous atom ($S = 2$). Consequently a $Fe^{3+}-Fe^{3+}$ pair with $S_{12} = 5$ will have a larger magnetic moment than a $Fe^{2.5+}_3-Fe^{2.5+}_4$ pair with $S_{34} = 9/2$. The sign of the $A'/g_n\beta_n$ values depends on whether these magnetic moments are aligned parallel or antiparallel to the applied field. From the observed data it follows, however, $A'_{12}/g_n\beta_n > 0$ for $Fe^{3+}_{(1)}-Fe^{3+}_{(2)}$ and $A'_{34}/g_n\beta_n < 0$ for $Fe^{2.5+}_{(3)}-Fe^{2.5+}_{(4)}$ (Table 7), and therefore we are concerned with the situation that the magnetic moment of the mixed-valence pair is larger than that of the $Fe^{3+}_{(1)}-Fe^{3+}_{(2)}$ pair. This situation is explained when the $Fe^{3+}_{(1)}-Fe^{3+}_{(2)}$ pair exhibits $S_{12} = 4$ instead of $S_{12} = 5$ [130, 131]. With this condition the reevaluation of the spin-coupling model for $[Fe_4S_4]^{3+}$ with $S_1 = S_2 = S_3 = 5/2$, $S_4 = 2$, $S_{12} = 4$, $S_{34} = 9/2$, $S_t = 1/2$ yields

$$\tilde{g}' = -(4/3)\tilde{g}_1 - (4/3)\tilde{g}_2 + (55/27)\tilde{g}_3 + (44/27)\tilde{g}_4 \tag{59}$$

and accordingly

$$\tilde{A}'_1 = -(4/3)\tilde{A}_1, \quad \tilde{A}'_2 = -(4/3)\tilde{A}_2, \quad \tilde{A}'_3 = (55/27)\tilde{A}_3,$$
$$\tilde{A}'_4 = (44/27)\tilde{A}_4. \tag{60}$$

Using the local values for Fe^{3+} and Fe^{2+} discussed before it is obvious that $g'_{av} > 2$, $A'_{12} > 0$ and $A'_{34} < 0$, in qualitative agreement with observation. Even quantitatively the estimated values

$$\tilde{A}'_{12}/g_n\beta_n \sim (+19\ T, +19\ T, +19\ T)$$
$$\tilde{A}'_{34}/g_n\beta_n \sim (-28.5\ T, -24.7\ T, -18.6\ T)$$

appear to be acceptable, when compared with those of Table 7.

5.4.1.5 Multiple Clusters

Several examples have been described in the literature where proteins contain more than one of the iron-sulfur clusters discussed here. This imposes, of course, additional difficulties in the analysis of experimental data of such systems, (i) due to the loss of spectral resolution because of the superposition of several sub-spectra, and (ii) also due to possible inter-cluster interactions. In spite of these difficulties, the characterization and understanding of multi-component systems was possible by the complementary use of various spectroscopies such as EPR and Mössbauer spectroscopy.

Examples of this kind are the $2[Fe_4S_4]$ ferredoxin from C. pasteurianum and its Se-substituted form [123, 124], the various hydrogenases [132], which may contain $[Fe_3S_4]$ and $[Fe_4S_4]$ clusters simultaneously and additionally even Ni [133], and E. coli sulfite reductase which exhibits exchange coupling between siroheme and a $[Fe_4S_4]$ cluster [134].

5.4.2 Linear and Triangular Iron-Containing Trinuclear Clusters

The spin-dependent part of the interaction between three paramagnetic sites 1, 2, 3 is expressed by the superexchange Hamiltonian

$$\mathscr{H}_{ex} = -J_{12}\vec{S}_1\cdot\vec{S}_2 - J_{23}\vec{S}_2\cdot\vec{S}_3 - J_{13}\vec{S}_1\cdot\vec{S}_3. \tag{61}$$

The eigenvalue problem for the general case of three different J values can be solved by diagonalizing $\mathscr{H}_{ex}$ on the basis of product states of individual sites $|S_1 M_1\rangle|S_2 M_2\rangle|S_3 M_3\rangle$ [135]. An example for the application of this formalism is provided by the magnetic susceptibility, Mössbauer [136], and molecular orbital [137] study of linear antiferromagnetically coupled clusters $[Fe_3S_4(SR)_4]^{3-}$, R = Et, Ph.

In the case of highly symmetric triangular bridged metal clusters as in $[Fe_3S_4]^{1+}$ from oxidized ferredoxin II [138], in $Fe_3O(SO_4)_6 \cdot 3H_2O$ [139], or in related compounds [140], the three coupling constants in Eq. (61) are equal.

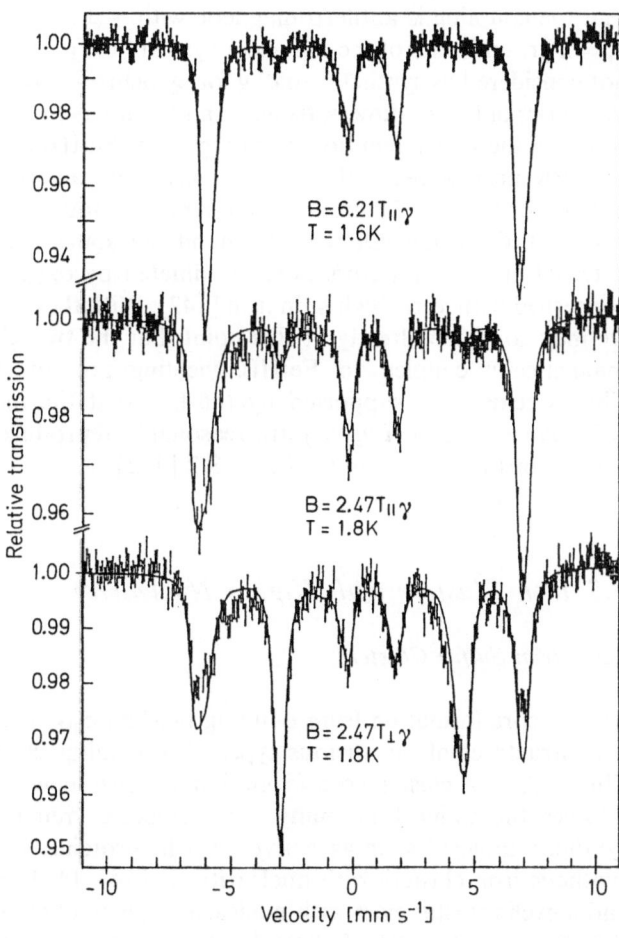

Fig. 42. Experimental Mössbauer spectrum of $\{Cu(mesalen)\}_2$ Fe(acac) $(NO_3)_2$ at temperatures and applied fields as indicated. The *solid lines* are spin-Hamiltonian simulations with $S_t = 3/2$. Taken from [142]

An even simpler example is the linear spin-coupled Cu–Fe–Cu unit in $\{Cu(mesalen)\}_2 Fe(acac)(NO_3)_2$, the exchange Hamiltonian of which, neglecting copper–copper interaction, is represented by

$$\mathscr{H}_{ex} = -J(\vec{S}_{Cu1} \cdot \vec{S}_{Fe} + \vec{S}_{Cu2} \cdot \vec{S}_{Fe}). \tag{62}$$

The study of spin-coupled iron–copper complexes is of interest because of the synergism of Fe(III) and Cu(II) in cytochrome oxidase, the respiratory enzyme that catalytically reduces dioxygen to water with concomitant release of energy. This enzyme contains four metal centers (two irons and two coppers) per functioning unit. The active or oxygen binding site is found to be a high-spin ferric heme (S = 5/2) paired with a cuprous ion (S = 1/2), EPR undetectable because of strong mutual coupling. In a magnetochemical study it was shown that the coupling is antiferromagnetic with $-J > 200 \text{ cm}^{-1}$ giving a resulting $S_t = 2$ ground state of the dinuclear system [141]. The linear Cu–Fe–Cu unit is not considered as synthetic analog for cytochrome oxidase, however, it belongs to a series of hetero dimers, trimers and tetramers [142], which are potential for studying the above mentioned synergism of Fe(III) and Cu(II).

EPR measurements of $\{Cu(mesalen)\}_2 Fe(acac)(NO_3)_2$ provide finger-print spectra for the $|\pm 1/2\rangle$ ground-state Kramers doublet of $S_t = 3/2$ with g-values of 4.68, 3.26 and 1.96. On the other hand, the isomer shift $\delta(\alpha\text{-Fe}) = 0.60 \text{ mms}^{-1}$ detected at 4.2 K for iron, hexacoordinated by oxygen, indicates that we are concerned with ferric high-spin iron [142]. With these two pieces of information we may conclude already at this point that the two Cu(II) ions are antiferromagnetically coupled to Fe(III), yielding ground state $|S_t = 3/2, \pm 1/2\rangle$. This conclusion is supported by the fact that the measured field-dependent Mössbauer spectra (Fig. 42) are reasonably reproduced by spin-Hamiltonian simulations using system spin $S_t = 3/2$ [142].

5.5 Iron Clusters with Higher Nuclearity

5.5.1 Iron Sulfur Clusters

Cluster-core formation from building blocks Fe_2S_2 is readily applicable to the core structures of the various types of iron-sulfur clusters discussed in [143]. This ability of cluster core formation suggests new synthetic targets and reinforces the point that many new structures remain to be discovered, as synthetic molecules or as active sites in proteins. Fascinating examples for synthetic iron clusters with nuclearity six [143–145], seven [146], eight [147], and a cyclic cluster even with nuclearity eighteen [143] have been described in the literature. Several of them have been the subject for a combined EPR, Mössbauer and magnetic susceptibility investigation [148].

5.5.2 Iron Oxo/Hydroxo Clusters

In recent years a steadily increasing number of oxo/hydroxo-bridged clusters of iron(III) have been synthesized and structurally characterized. Complexes of nuclearity Fe_3, Fe_4, Fe_6, Fe_8, Fe_{11}, and even higher, have been isolated, and their spectroscopic and magnetic properties have been investigated in considerable detail [149–153]. Interest in this chemistry of biominerals has been stimulated by the fascinating problem of the uptake and release of iron in the iron storage proteins ferritin [154–156] and haemosiderin [157, 158]. Another interesting problem associated with biomineralization is the iron uptake by teeth and bones [159, 160] and the magnetic-grain formation by magnetotactic bacteria [161]. Here we will concentrate on ferritin and haemosiderin and discuss as an example for EPR and Mössbauer applications the superparamagnetic behavior of iron aggregates in these biomolecules.

Ferritin is the iron storage protein in living organisms, and has been isolated from mammals, bacteria and plants [156]. When iron concentration is high, either as a result of disease or other factors, another iron storage material, haemosiderin, may be also present [157]. Both ferritin and haemosiderin, are composed of a protein shell which encloses an iron(III)-containing crystalline core. The inner diameter of the protein shell in ferritin is about 8 nm and can be filled to varying degrees with an inorganic "hydrous ferric oxide-phosphate" complex [162], up to a maximum of about 4500 iron atoms in a single ferritin molecule. Unlike ferritin, haemosiderin has only a partial polypeptide shell, however, its iron-containing core is considered to be very similar to that of ferritin. Mössbauer studies of iron-rich ferritin and haemosiderin show at 4.2 K a well resolved magnetic hyperfine pattern, which by increasing temperature, changes into a quadrupole doublet [163] (Fig. 43). The transition temperature, which is also termed blocking temperature (T_B), depends on the size of the iron core. This behaviour occurs in small particles of magnetically ordered materials and is called superparamagnetism.

The magnetization direction in each superparamagnetic particle changes between the easy magnetic axes at a rate

$$f = f_0 \exp(-KV/kT), \tag{63}$$

where K is the magnetic anisotropy constant, V is the volume of the particle, k is the Boltzmann constant and T the temperature. f_0 is a constant of about 10^9 s^{-1} [164]. The magnetic spectrum collapses to a doublet at T_B, at which f is equal to the nuclear Larmor frequency in the magnetic hyperfine field. In ferritin the blocking temperature for particles with a diameter of about 7 nm is 50 K. This yields the relation between the diameter of the iron aggregates and the blocking temperature as $d = (6860 \, T_B)^{1/3}$, for particles with the same magnetic anisotropy constant K. It was argued that this anisotropy constant is the same in various ferritins and haemosiderins [162]. Assuming that both proteins have the same K values, the higher blocking temperature for haemosiderin (Fig. 43)

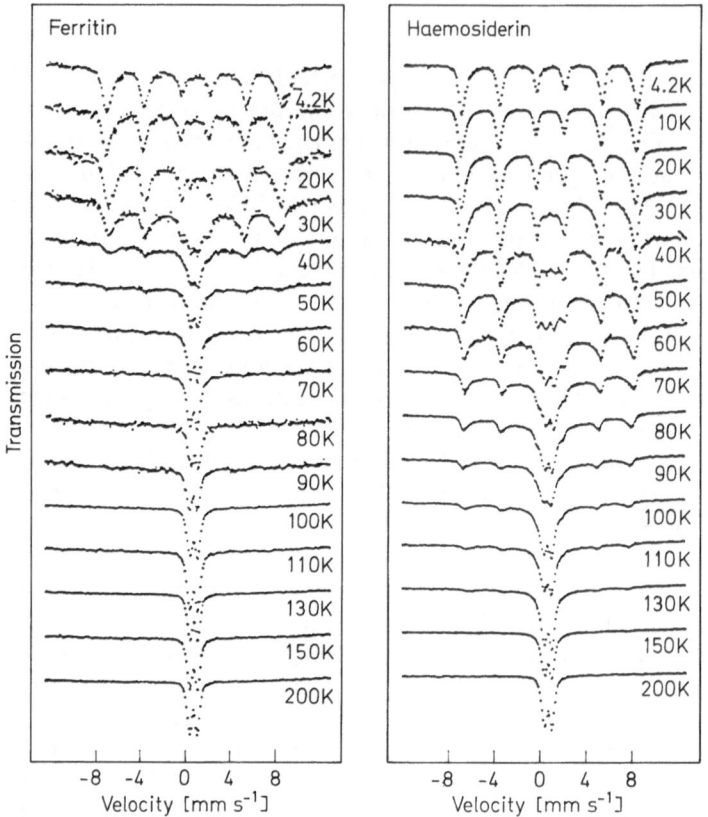

Fig. 43. Experimental Mössbauer spectra of human ferritin and haemosiderin at various temperatures as indicated. Taken from [163]

would mean that the haemosiderin cores are larger than those of ferritin. However, contrary to this, electron microscopy shows that haemosiderin cores are markedly smaller on average than those of ferritin [163]. The superparamagnetic fluctuations observed by EPR spectroscopy for ferritin and haemosiderin [158] are in qualitative agreement with this latter observation.

For estimates of particle sizes from temperature-dependent Mössbauer spectra according to Eq. (63) it is therefore desirable to use anisotropy constants which are derived independently. Temperature-dependent EPR measurements provide this possibility. EPR line widths of superparamagnetic particles decrease drastically with increasing temperature, because anisotropic magnetocrystalline fields seen by the EPR experiment are averaged out within the time window of $\sim 10^{-10}$ s for X-band EPR. In favoured cases the anisotropy constant K may be obtained from the narrowing of EPR lines [165] as demonstrated for small iron particles in zeolites [166].

5.6 Spin-Coupled Systems Containing Only One Metal Site

The interaction of transition metals with nonmetallic paramagnetic species (i.e. radicals, O_2 etc.) in active protein sites, notably of exchange and spin–dipolar type, can be studied spectroscopically. Examples are the metal-radical spin coupling in oxoferryl porphyrin cation radical complexes and the iron–dioxygen interaction in Fe(II)-substituted horse liver alcohol dehydrogenase (HLADH).

In HLADH the catalytic Zn(II) ion was replaced by Fe(II), providing a novel iron protein with the unusual coordination of Fe(II) to two cysteine residues, one histidine residue and water. The electronic structure of iron in this system was characterized by Mössbauer spectroscopy at various temperatures as well as applied fields and analyzed in terms of the spin-Hamiltonian formalism [167]. The novelty found in this system is an unusually weak spin coupling ($|J| < 0.1$ cm^{-1}) of a paramagnet ($S = 1$) with iron ($S = 2$). From EPR and biochemical studies it is concluded that the corresponding chemical species is triplet oxygen (O_2). The quantitative determination of the very small coupling energy was possible utilizing the competition between Zeeman interaction and spin coupling at weak magnetic fields and low temperature.

In the following we discuss the iron-radical interaction in oxoferryl porphyrin, detected by Mössbauer and EPR spectroscopy, in more detail. High-valency oxoheme complexes, formally two-electron oxidized above the ferric state, have been of considerable interest because of their importance in the chemistry of catalase and peroxidase enzymes [168, 169], and because they play a putative role in monooxygen transfer to organic substrates by cytochrome P450 [170, 171]. Efforts to model biochemical function and to characterize better the electronic structure of these intermediates have resulted in the synthesis of oxoheme complexes in the same oxidation state as the enzymatic transients [172, 173] and in mimicking some of the biochemical functions of cytochrome P450 [173, 174].

The Mössbauer spectra of ^{57}Fe-enriched iron(IV) porphyrin radical complexes, recorded at temperatures above 50 K in zero-magnetic field, are plain quadrupole doublets. Internal fields average to zero due to fast spin relaxation. Table 8 summarizes the values from various complexes with different porphyrins, counter ions and solvents. The ferryl oxidation state of iron in these complexes is suggested from the isomer shifts which is typical for complexes containing the $[FeO]^{3+}$ unit.

For the analysis of the magnetic Mössbauer spectra of iron(IV) ($S = 1$) porphyrin radical ($S' = 1/2$) systems [175, 176] (Fig. 44) the local spin Hamiltonian includes iron-radical spin coupling. The spin-coupling tensor $\tilde{J}$ decompose into an isotropic part $1\tilde{J}$, and a traceless anisotropic contribution $\tilde{J}^d$, originating from exchange interaction and spin–dipolar interaction, respectively, between iron and cation radical. At the limit of strong exchange interaction, where isotropic exchange dominates spin–dipolar, zero-field and Zeeman interactions, the local-spin Hamiltonian may be replaced in the

Table 8. Mössbauer parameters of porphyrin complexes containing Fe(IV)

Compound	T(K)/matrix	$\delta(\alpha\text{-Fe})$ (mms^{-1})	ΔE_Q (mms^{-1})	Ref.
[(TMP)Fe=O]$^+$(Cl$^-$)	77/toluene/methanol (4:1)	0.06	1.62	[175]
	4.2/toluene/methanol (4:1)	0.08	1.62	
[(TMP)Fe=O]$^+$(OSO$_2$CF$_3$)	4.2/toluene	0.08	1.62	[176]
[TPP(2,6-Cl)Fe=O]$^+$(OSO$_2$CF$_3$)	4.2/CH$_3$CN	0.08	1.80	
(TPP)Fe=O/1-methylimidazole	4.2/toluene	0.11	1.26	b
(TPP)Fe=O/pyridine	4.2/toluene	0.10	1.56	
(TP$_{piv}$P)Fe=O/tetrahydrofuran	4.2/tetrahydrofuran	0.12	2.20	c
(TP$_{piv}$P)Fe=O/1-methylimidazole	4.2/tetrahydrofuran	0.109	1.372	
(TMP)Fe=O	nd/nd	0.04	2.3	d
TPP(2,6-Cl)Fe=O/tetrahydrofuran	4.2/tetrahydrofuran	0.09	2.08	e
TPP(2,6-Cl)Fe=O/dimethylformamide	4.2/dimethylformamide	0.09	1.81	
TPP(2,6-Cl)Fe=O/1-methylimidazole	4.2/tetrahydrofuran	0.07	1.35	f
Compound ES cytochrome c peroxidase	4.2/H$_2$O	0.05	1.55	g
(TMP)Fe(OCH$_3$)$_2$	4.2/toluene	-0.022	2.17	
Horse-radish peroxidase I	$-$^{a}/glycerol/H$_2$O(1:1)	0.08	1.25	[179]
Horse-radish peroxidase II	77/H$_2$O	0.03	1.36	h
Chloroperoxidase I	4.2/peracetic acid	0.15	1.02	[102]

[a] Parameters derived from simulations of temperature-dependent and field-dependent Mössbauer spectra.
[b] Simonneaux et al. (1982) Biochim. Biophys. Acta 716:1.
[c] Schappacher et al. (1985) J. Am. Chem. Soc. 107:3736.
[d] Groves et al. (1986) Inorg. Chem. 25:123.
[e] Gold et al. (1988) J. Am. Chem. Soc. 110:5756.
[f] Lang et al. (1976) Biochim. Biophys. Acta, 451:250.
[g] Groves et al. (1985) J. Am. Chem. Soc. 107:354.
[h] Moss et al. (1969) Biochemistry 8:4159

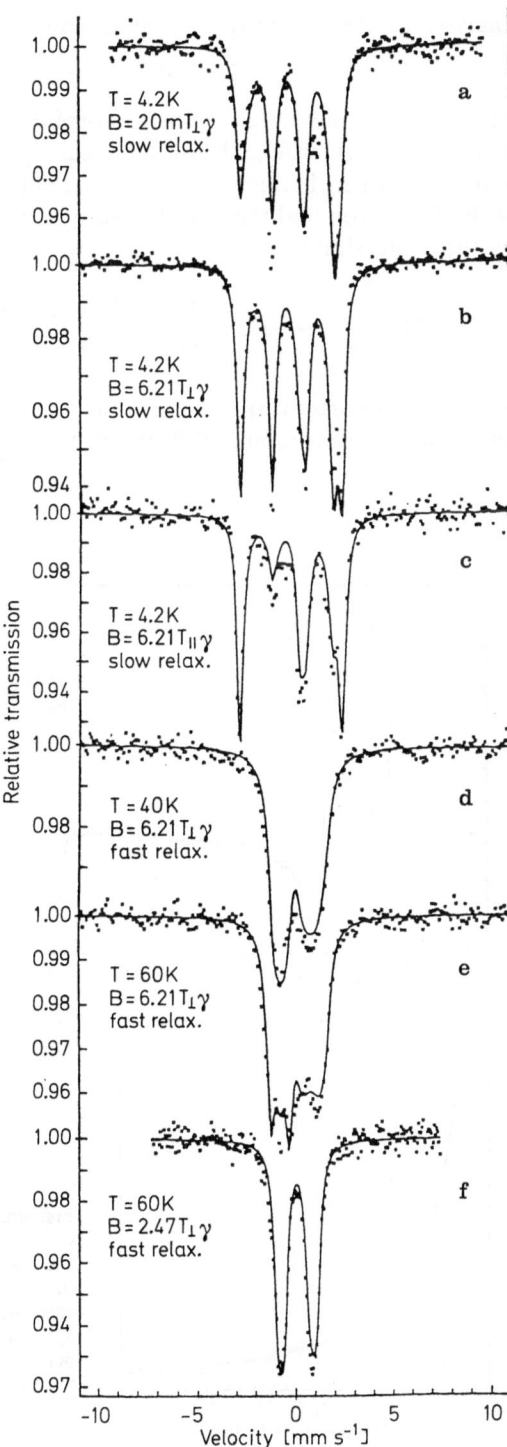

Fig. 44. Experimental Mössbauer spectra of $[(TMP)Fe=O]^+$ (Cl^-) at various temperatures and applied fields as indicated. The *solid lines* are simulated spectra with the parameters given in Table 8. Taken from [9]

analyses by the system-spin Hamiltonian

$$\mathscr{H}_e = D_t[S_{tz}^2 - (1/3)S_t(S_t + 1) + (E_t/D_t)(S_{tx}^2 - S_{ty}^2)] + \beta g\vec{S}_t \cdot \vec{B}, \quad (64)$$

with $\vec{S}_t$ being the system-spin (total-spin) operator $\vec{S} + \vec{S}'$. For parallel spin coupling of ferryl iron and cation radical ($S_t = 3/2$), the zero-field parameters D_t and E_t, occurring in this expression, are related to the local zero-field parameters of the ferryl iron and the anisotropic spin-coupling components J_x^d, J_y^d, and J_z^d, by the expression [177]

$$D_t = \frac{1}{3}D - \frac{1}{2}J_z^d, \qquad \frac{E_t}{D_t} = \frac{2E - J_x^d - J_y^d}{2D - 3J_z^d}. \quad (65)$$

In Eq. (65), it is assumed that the principal axes of $\tilde{D}$ and $\tilde{J}^d$ tensors coincide. For antiparallel spin coupling of ferryl iron and cation radical ($S_t = 1/2$) the para-

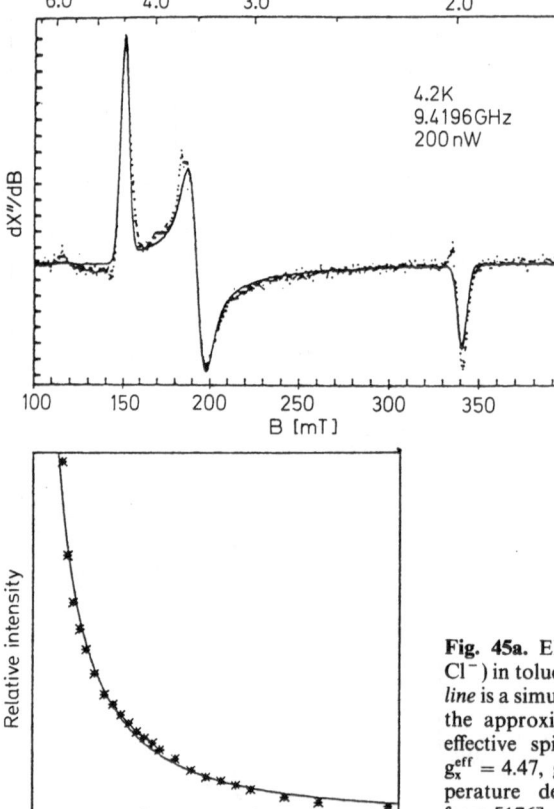

Fig. 45a. EPR spectrum of $[(TMP) Fe = O]^+$ Cl^-) in toluene/methanol (4:1) at 4.2 K. The *solid line* is a simulation of a powder spectrum based on the approximation of a Kramers doublet with effective spin $S^{eff} = 1/2$ and effective g-values $g_x^{eff} = 4.47$, $g_y^{eff} = 3.50$, and $g_z^{eff} = 1.98$. **(b)** Temperature dependence of EPR signal. Taken from [176]

meters D_t and E_t are meaningless so that Eq. (64) contains the Zeeman term only. After diagonalization of the spin Hamiltonian, spin-expectation values of the iron spin are calculated. In the limit of strong coupling, the local iron values $\langle \vec{S} \rangle$ follow from projection of total spins: $\langle \vec{S} \rangle = (2/3) \langle \vec{S}_t \rangle$ for $S_t = 3/2$, and $\langle \vec{S} \rangle = (4/3) \langle \vec{S}_t \rangle$ for $S_t = 1/2$.

A representative example for a spin-coupled iron-radical pair is $[(TMP)Fe=O]^+(Cl^-)$, the EPR spectrum of which is shown in Fig. 45a. The resonances at effective g-values of $g^{eff} \sim 4$ and $g^{eff} = 2$ are characteristic of a $|S_t = 3/2, \pm 1/2\rangle$ Kramers doublet, which indicates that iron and radical spins in $[(TMP)Fe=O]^+(Cl^-)$ are strongly coupled, yielding total spin $S_t = 3/2$ in its ground state. The temperature dependence of the EPR line intensity (Fig. 45b) reveals that the $|\pm 1/2\rangle$ Kramers doublet is lowest in energy within the $S_t = 3/2$ multiplet, indicating that the zero-field parameter D_t is positive. By fitting the EPR line intensities with appropriate Boltzmann factors the zero-field parameter D_t is found to be in the range 6–$12\ cm^{-1}$. This agrees with the value $+6.2 \pm 2\ cm^{-1}$ deduced from simulations of Mössbauer spectra of $[(TMP)Fe=O]^+(Cl^-)$ [175, 176]. In these simulations (Fig. 44), the local zero-field parameter $D = 3D_t = 18.6\ cm^{-1}$ was used. The EPR spectrum in Fig. 5.19a exhibits considerable rhombic splitting $\Delta g = |g_x^{eff} - g_y^{eff}|$ of the $g_{x,y}^{eff}$ signals. From the first-order perturbation expression $\Delta g = 12\ E_t/D_t$, the experimental value $\Delta g = 0.97$ yields the rhombicity $E_t/D_t = 0.08$. The rhombicities of the TMP complexes, $[(TMP)Fe=O]^+$, are weakly dependent on solvent, and independent of the counter ions Cl^- or $OSO_2CF_3^-$. These observations suggest that the counter ions are not coordinated to iron in TMP complexes, in accordance with the conclusions drawn from an EXAFS study of $[(TMP)Fe=O]^+(Cl^-)$ [178]. Simulations of Mössbauer spectra of the strongly coupled complexes $[(TMP)Fe=O]^+(Cl^-)$ and $[(TMP)Fe=O]^+(OSO_2CF_3^-)$ turned out to be insensitive to values of E/D in the range 0–0.08.

The magnetic Mössbauer spectra of $[(TMP)Fe=O]^+(Cl^-)$ in toluene/methanol (4:1) were analyzed within the system-spin Hamiltonian ($S_t = 3/2$) [175] and also within the spin Hamiltonian (Fig. 44) [176] which explicitly includes iron-radical interaction. The latter provides information about the magnitude of exchange coupling, which is $J > 80\ cm^{-1}$ in the present case. We notice that the final values of the spin-Hamiltonian parameters (Table 9), which are needed to simulate a series of temperature- and field-dependent spectra consistently, did not result from the spin-Hamiltonian analysis alone. The parameters ΔE_Q, δ and Γ can be determined from zero-field Mössbauer spectra, and D, E/D are from EPR data. Thus, only the hyperfine-coupling components A_x, A_y, A_z, and the exchange-coupling parameter J are left as adjustable parameters in the simulations of the magnetic Mössbauer spectra.

In Table 9 we have summarized the spin-Hamiltonian parameters of several oxoferryl porphyrin cation radical complexes. The two prominent features of these parameters are (i) the relatively large zero-field splittings and (ii) the varying sign and magnitude of exchange coupling constants. Both features will be discussed in the following.

Table 9. Spin-Hamiltonian parameters of oxoferryl porphyrin cation radical complexes

Compound/matrix	D (cm^{-1})	E/D	ΔE_Q (mms^{-1})	η	$\dfrac{A_{x,y}(T)}{g_n\beta_n}$	$\dfrac{A_z(T)}{g_n\beta_n}$	$\delta(\alpha\text{-Fe})$ (mms^{-1})	Γ (mms^{-1})	J (cm^{-1})	Ref.
[(TMP)Fe=O]$^+$(Cl$^-$)/toluene/methanol	18.6	0	+1.62	0	−22.1	−10.0	0.06	0.50	∞	[175]
	18.6	0–0.08ᵃ	+1.62	0	−23.1	−10.0	0.08	0.40	>80	[176]
[(TMP)Fe=O]$^+$(OSO$_2$CF$_3$)/toluene	18.6	0–0.06ᵃ	+1.62	0	−23.1	−10.0	0.08	0.40	>80	[176]
Horseradish peroxidase I/glycerol/H$_2$O	22.2	0	+1.25	0	−17.0	−6.0	0.08	—	3	[66]
	25.7	0	+1.25	0	−19.3	−6.0	0.08	0.31	3	[179]
Chloroperoxidase I/peracetic acid	36.1	0.03	+1.02	0	−20.2	−6.0	0.15	—	−36	[102]

ᵃ Mössbauer simulations are insensitive to variations of E/D in the given range. The upper limit of the range is determined by EPR measurements

The contributions to D and E/D, originating from local zero-field inter-actions and from spin-dipolar iron-radical interactions, according to Eq. (65), cannot be distinguished. For example, substitution of local zero-field parameters $D = 24 \, \mathrm{cm}^{-1}$, $E = 0$, and of spin–dipolar parameters $J_x^d = -3.5 \, \mathrm{cm}^{-1}$, $J_y^d = -0.5 \, \mathrm{cm}^{-1}$, $J_z^d = 4 \, \mathrm{cm}^{-1}$, in Eq. (65), would yield $D_t = 6 \, \mathrm{cm}^{-1}$ and $E_t/D_t = 0.08$, in agreement with EPR and Mössbauer data of $[(\mathrm{TMP})\mathrm{Fe}{=}\mathrm{O}]^+$ (Cl^-). The values taken for the spin–dipolar coupling seem rather large, though similar values have been suggested for HRP I [179] and CPO I [102]. J^d, which is typically of the order $g^2\beta^2 r^{-3}$, with r denoting the separation between the coupled spins, is for a nearest-neighbour distance of 0.2 nm equal to only $0.2 \, \mathrm{cm}^{-1}$. This value is confirmed by more sophisticated theoretical estimates of J^d, obtained by numerical integration of the spin-Hamiltonian, using reasonable radial dependences of the unpaired-electron orbitals and adopting a radical orbital centered at 0.2 nm from iron [180]. The rhombicity is, thus, most likely to be a consequence of the non-axial zero-field properties of the iron ion.

Although the general formulation of the electronic structure of the metal site in oxoferryl porphyrin cation radical complexes is similar, the iron–radical interaction in synthetic analogs, CPO I and HRP I shows considerable varia-tion. The cation radical is ferromagnetically coupled to the ferryl iron in $[(\mathrm{TMP})\mathrm{Fe}{=}\mathrm{O}]^+(\mathrm{Cl}^-)$ [175, 176], antiferromagnetically coupled in CPO I [102], and only weakly coupled in HRP I [66, 179] (Table 9). The poten-tial second axial ligand in these porphyrins is probably nonexisting in $[(\mathrm{TMP})\mathrm{Fe}{=}\mathrm{O}]^+ \mathrm{Cl}$, while it is sulfur (cysteine) in CPO I and nitrogen (histidine) in HRP I. Thus, the differences observed in spin-coupling are likely associated with structural differences due to the second axial ligand, and probably also with structural differences in the porphyrin moiety. Metal porphyrin cation radical complexes often present a saddle-shaped core [181–185]. In these structures, the pyrrole rings are displaced alternately above and below the mean plane of the core, and the meso carbons are found to be in, or nearly in, this mean plane. The highest possible symmetry of a saddle-shaped core is D_{2d}. Assuming that a saddle-shaped core is also characteristic of the oxoferryl porphyrin cation radical complexes under study, the dissymmetrical axial ligation of iron yields C_{2v} as the highest possible symmetry of the metal site, consistent with the rhombic distortion found in the EPR analysis of $[(\mathrm{TMP})\mathrm{Fe}{=}\mathrm{O}]^+(\mathrm{Cl}^-)$ (Fig. 45a).

The structurally-related trends in sign and magnitude of the exchange interactions between the iron and radical electrons may be rationalized in terms of a simple model [19, 74, 104, 186]. The sign of J depends, in such a description, on the overlap properties of the radical-electron and 3d-electron orbitals. In the case of the orbitals being orthogonal, the exchange coupling constants have ferromagnetic sign, and can be expressed in terms of exchange integrals between the radical and 3d orbitals. The size of J then depends on the overlap-charge density between these orbitals. When the radical and 3d orbitals are non-orthogonal, the expression for J becomes more complicated, involving terms depending on overlap and nuclear attraction integrals. In this case the exchange coupling tends to favour antiferromagnetic coupling. The relative importance of the ferromagnetic and antiferromagnetic contributions to the exchange-coupling constants in paramagnetic complexes is dependent sensitively on their structures [104]. In $[(TMP)Fe=O]^+(Cl^-)$ we are in the orthogonal regime. In HRP I there is an accidental cancellation of ferromagnetic and antiferromagnetic contributions to J, whereas in CPO I the antiferromagnetic contributions are dominant. In C_{2v} symmetry, the singly occupied 3d orbitals of ferryl iron, d_{xz} and d_{yz}, have b_1 and b_2 symmetry, respectively. The radical electron orbitals, a_{1u} or a_{2u} in D_{4h} symmetry, become in C_{2v} symmetry a_2 and a_1, respectively, and are strictly orthogonal to the 3d orbitals, and thus should lead to ferromagnetic exchange coupling. The antiferromagnetic coupling in CPO I indicates either a symmetry lower than C_{2v} or that the d_{xy} orbital is not the most stable 3d orbital of the iron atom. Significant changes in the exchange and spin–dipolar couplings may arise when the radical electron is transferred to another orbital, under certain structural conditions, which may be associated with porphyrin shape and axial ligation. Transition from an a_{2u}-like orbital, with considerable electron density at the pyrrole nitrogen centers, to an a_{1u}-like orbital, which vanishes at these centres, will inevitably lead to a reduction of the coupling constants $|J|$ and $|J^d|$. Which mechanism is responsible for the reduction of the coupling in HRP I and for the sign-reversal of J in CPO I relative to the coupling in $[(TMP)Fe=O]^+(Cl^-)$ is not yet clear.

Acknowledgements: We wish to acknowledge the fruitful cooperation with Professors A. Gold (Chapel Hill), A. Kostikas (Athens), R. Weiss (Strasbourg), K. Wieghardt (Bochum), M. Zeppezauer (Saarbrücken) and their groups, with whom many of the results which are described in this article have been obtained. We are very much indebted to our coworkers at the Institut für Physik (Lübeck), and there especially to Dr. X.-Q. Ding. We gratefully acknowledge the financial support by the Deutsche Forschungsgemeinschaft and by Stiftung Volkswagenwerk.

6 References

1. Lang G (1970) Quart Rev Biophys 3: 1
2. Münck E, Zimmermann R (1976) In: Gruverman IJ, Seidel CW (eds) Mössbauer effect methodology, vol 10 Plenum, New York p 191
3. Münck E (1978) In: Fleischer S, Pa)cker L (eds) Biomembranes, part E: Biological oxidations, Specialized techniques Academic, New York p 346 (Methods in enzymology, vol LIV)
4. Huynh BH, Kent TA (1984) In: Eichhorn GL, Marzilli LG (eds) Advances in inorganic biochemistry Elsevier, New York p 163 (Advances in inorganic biochemistry, vol 6)
5. Debrunner PG (1989) In: Lever ABP, Gray HB (eds) Iron porphyrins, part III, VCH Publishers, New York p 137
6. Abragam A, Pryce MHL (1951) Proc Roy Soc (London) 205A: 135
7. Abragam A, Bleaney B (1970) Electron paramagnetic resonance of transition ions Clarendon, Oxford
8. Gismelseed A, Bominaar EL, Bill E, Trautwein AX, Weiss R (1990) (in preparation)
9. Griffith JS (1961) The theory of transition ions Cambridge University Press, Cambridge
10. Gismelseed A, Bominaar EL, Bill E, Trautwein AX, Winkler H, Nasri H, Doppelt P, Mandon D, Fischer J, Weiss R (1990) Inorg Chem 29: 2741

11. Wickman HH, Klein MP, Shirley DA (1966) Phys Rev 152: 345
12. Blume M (1968) Phys Rev 174: 351
13. Clauser MJ, Blume M (1971) Phys Rev B3: 583
14. Winkler H, Schulz C, Debrunner PG (1979) Phys Lett 69A: 360
15. Winkler H, Bill E, Trautwein AX, Kostikas A, Simopoulos A, Terzis A (1988) J Chem Phys 89: 732
16. Dickson DPE, Berg FJ (eds) (1986) Mössbauer spectroscopy Cambridge University Press, Cambridge
17. Steven JG, Gettys WL (1978) in: Shenoy GK, Wagner FE (eds) Mössbauer isomer shift. North Holland, Amsterdam p 901
18. Kittel C (1976) Introduction to solid state physics, 5th edn John Wiley, New York
19. van Vleck JH (1932) The theory of electric and magnetic susceptibilities Oxford University Press, Oxford
20. Connor CC (1982) Prog Inorg Chem 29: 203
21. English DR, Hendrickson DN, Suslick KS (1984) J Am Chem Soc 106: 7258
22. Griffith JS (1957) Nature 180: 30
23. Kotani M (1961) Prog Theoret Phys Suppl 17: 4
24. Blumberg WE, Peisach J (1971) In: Chance B, Yonetani T, Mildvan AS (eds) Probes of structure and function of macromolecules and membranes Academic, New York (vol 2)
25. Palmer G (1983) In: Lever ABP, Gray HB (eds) Physical bioinorganic chemistry series, Addison Wesley, London (Iron porphyrins, part II) p 43
26. Lang G, Marshall W (1966) Proc Phys Soc 87: 3
27. Lang G, Herbert D, Yonetani T (1968) J Chem Phys 49: 944
28. Rhynard D, Lang G, Spartalian K, Yonetani T (1979) J Chem Phys 71: 3715
29. Mitra S (1983) In: Lever ABP, Gray HB (eds) Physical bioinorganic chemistry series, Addison Wesley, London (Iron porphyrins, part II) p 1
30. Mashiko T, Kastner ME, Spartalian K, Scheidt WR, Reed CA (1978) J Am Chem Soc. 100: 6354
31. Peisach J, Blumberg WE, Lode ED, Coon MJ (1971) J Biol Chem 246: 5877
32. Miyamoto T, Ogino N, Yamamoto S, Hayaishi DJ (1976) Biol Chem 251: 2629
33. Kulmacz RJ, Lands WEM (1984) J Biol Chem 259: 6358
34. Karthein R, Nastainczyk W, Ruf HH (1987) Eur J Biochem 166: 173
35. Alpert Y, Conder Y, Tuckender J, Thome H (1973) Biochem Biophys Acta 322: 34
36. Yim MB, Kuo LC, Makinen MW (1982) J Magn Res 46: 247
37. Colvin JT, Rutter R, Stapelton HJ, Hager LP (1983) Biophys J 41: 105
38. Orbach R, Stapleton HJ (1972) In: Geschwind S (ed) Electron paramagnetic resonances, Plenum Press, New York, p 121
39. Slappendel S, Veldink GA, Vliegenhardt JFG, Aasa R, Malmström B (1980) Biochem Biophys Acta 642: 30
40. Karthein R (1987) Untersuchung von Struktur und Mechanismus des Hämproteins Prostaglandin H Synthase mittels Elektronen-spinresonanz und Mößbauer-spektroskopie. Thesis, Universität des Saarlandes, Saarbrücken
41. Greenwood NN, Gibb TC (1971) Mössbauer spectroscopy Chapman and Hall, London
42. Bernhardt FH, Heymann E, Traylor PS (1978) Eur J Biochem 92: 209
43. Bill E, Bernhardt, FH, Trautwein AX, Winkler H (1985) Eur J Biochem 147: 177
44. Twilfer H, Bernhardt FH, Gersonde K (1985) Eur J Biochem 147: 171
45. Marathe VR, Trautwein AX, Kostikas A (1980) J Phys Colloque 41, Cl: 315
46. Marathe VR, Trautwein AX (1983) In: Thosar BV, Iyengar PK (eds) Advances in Mössbauer spectroscopy, Elsevier, Amsterdam, p 398
47. Wickman HH, Trozzolo AM (1968) Inorg Chem 7: 63
48. Ganguli P, Marathe VR, Mitra S (1975) Inorg Chem 14: 970
49. Ganguli P, Hasselbach KM (1979) Z Naturforsch 34a: 1500
50. Niarchos D, Kostikas A, Simopoulos A, Coucouvanis D, Piltingsrud D, Coffman RE (1978) J Chem Phys 69: 4411
51. Wells FV, McCann SW, Wickman HH, Kessel SL, Hendrickson DN, Feltham RD (1981) Inorg Chem 21: 2306
52. Lovenberg W (ed) (1973) Iron sulfur proteins Academic, New York (Vol I–III)
53. Fee JA, Findling KL, Yoshida T, Hille R, Tarr GE, Hearshen DO, Dunham WR, Day EP, Kent TA, Münck E (1984) J Biol Chem 259: 124

54. Cline JF, Hoffman BM, Mims WB, LaHaie E, Ballou DP, Fee JA (1985) J Biol Chem 260: 3251
55. Moura I, Huynh BH, Hausinger RR, LeGall J, Xavier AV, Münck E (1980) J Biol Chem 255: 2493
56. Schulz C, Debrunner PG (1976) J Phys Colloque 37, C6: 153
57. Aasa R, Vänngard T (1975) J Magn Resonance 19: 308
58. Yang AS, Gaffney BJ (1987) Biophys J 51: 55
59. Berg JM, Holm RH (1982) In: Spiro TG (ed) Iron sulfur proteins Wiley, New York
60. Papaefthymiou V, Simopoulos A, Kostikas A, Coucouvanis D (1986) Hyperfine Interact 30: 337
61. Holm RH, Ibers A (1977) In: Lovenberg W (ed) Iron sulfur proteins Academic, New York (vol III)
62. Trautwein AX, Bill E, Bläs R, Lauer S, Winkler H, Kostikas A (1985) J Chem Phys 82: 3584
63. Winkler H, Bill E, Lauer S, Lauer U, Trautwein AX, Kostikas A, Papaefthymiou V, Bögge H, Müller A, Gerdau E, Gonser U (1985) J Chem Phys 82: 3595
64. Zimmermann R (1975) Nucl Instrum Methods 128: 537
65. Winkler H, Kostikas A, Petrouleas V, Simopoulos A, Trautwein AX (1986) Hyperfine Interact 29: 1347
66. Schulz CE, Rutter R, Sage JT, Debrunner PG, Sager LP (1984) Biochemistry 23: 4743
67. Scheidt WR, Gouterman M (1983) In: Lever ABP, Gray H (eds) Iron porphyrins, part I. Addison-Wesley, London
68. Maltempo MM, Moss TH (1976) Q Rev Biophys 9: 181
69. Maltempo MM, Moss TH, Spartalian K (1980) J Chem Phys 73: 2100
70. Scheidt WR, Osvath SR, Lee YJ, Reed CA, Shaevitz B, Gupta GP (1989) Inorg Chem 28: 1591
71. Schulz CE (1979) Mössbauer and EPR studies of iron-containing proteins Thesis, University of Illinois
72. Zener C (1951) Phys Rev 82: 403
73. Jonker GH, van Santen JH (1950) Physica 16: 337
74. Anderson PW, Hasegawa H (1955) Phys Rev 100: 675
75. Johnson CE, Elstner E, Gibson JF, Benfield G, Evans MCW, Hall DO (1968) Nature 220: 1291
76. Dunham WR, Bearden AJ, Salmeen IT, Palmer G, Sands RH, Orme-Johnson WH, Beinert H (1971) Biochim Biophys Acta 253: 134
77. Kessler H, Ly S (1981) J Sol State Chem 39: 22
78. Gillum WO, Frankel RB, Foner S, Holm RH (1976) Inorg Chem 15: 1095
79. Bominaar EL, Block R (1983) Physica 121 B + C: 109
80. Hatfield WE (1974) In: Interrante LV (ed) Extended interactions between metal ions in transition metal complexes, ACS Symposium Series No 5, American Chemical Society, Washington D.C., p 108
81. Kahn O, Charlot MF (1980) Nouv J Chim 4: 567
82. van Kalkeren G, Schmidt WW, Block R (1979) Physica 97 B + C: 315
83. Jansen L, Block R (1977) Physica 86–88 B: 1012
84. Noodleman L, Baerends EJ (1984) J Am Chem Soc 106: 2316
85. Bläs R, Guillin J, Bominaar EL, Grodzicki M, Marathe VR, Trautwein AX (1987) J Phys: At Mol Phys 20: 5627
86. Adler J, Ensling J Gütlich P, Bominaar EL, Guillin J, Trautwein AX (1988) Hyperf Interactions 42: 869
87. Noodleman L, Norman JG jr (1979) J Chem Phys 70: 4903
88. Petersson L, Cammack R, Rao KK (1980) Biochim Biophys Acta 622: 18
89. Palmer G (1973) In: Lovenberg W (ed) Iron sulfur proteins, Academic, New York, vol 2, p 285
90. Mason R, Zubieta JA (1973) Angew Chem 9: 390
91. Blumberg WE, Peisach J (1974) Archives Biochem Biophys 162: 502
92. Gibson JF, Hall DO, Thornley JHM, Whatley FR (1966) Proc Nat Acad Sci USA 56: 987
93. Bertrand P, Guigliarelli B, Gayda JP, Bearwood P, Gibson JF (1985) Biochim Biophys Acta 831: 261
94. Wilkins RG, Harrington PC (1983) Adv Inorg Biochem 5: 51
95. Antanaitis BC, Aisen P (1983) Adv Inorg Biochem 5: 111
96. Woodland MP, Dalton H (1984) J Biol Chem 259: 53
97. Debrunner PG, Hendrich MP, Jersey JD, Keough DT, Sage JT, Zerner B (1983) Biochim Biophys Acta 745: 103
98. Borovik AS, Murch BP, Que L jr, Papaefthymiou V, Münck E (1987) J Am Chem Soc 109: 7190

99. Suzuki M, Uehara A, Oshio H, Endo K, Yanaga M, Kida S, Saito K (1987) Bull Chem Soc Jpn 60: 3547
100. Hartman JR, Rardin RL, Chaudhuri P, Pohl K, Wieghardt K, Nuber B, Weiss J, Papaefthymiou GC, Frankel RB (1987) J Am Chem Soc 109: 7387
101. Kurtz PM jr (1990) Chem Rev 90: 585
102. Rutter R, Hager LP, Dhonau H, Hendrich M, Valentine M, Debrunner PG (1984) Biochem 23: 6809
103. Bläs R (1985) IEHT und SCC-Xα MO Rechnungen an Ferredoxin zur Bestimmung elektronischer Konfigurationen und Interpretation Spektroskopischer Daten Thesis, Universität des Saarlandes
104. Willet RD, Gatteschi D, Kahn O (eds) (1985) Magnetostructural Correlations in Exchange-Coupled Systems Reidel, Dodrecht
105. Bencini A, Gatteschi D (1990) Electron paramagnetic resonance of exchange-coupled systems Springer, Berlin Heidelberg New York
106. Snyder BS, Patterson GS, Abrahamson AJ, Holm RH (1989) J Am Chem Soc 111: 5214
107. Surerus KK, Münck E, Snyder BS, Holm RH (1989) J Am Chem Soc 111: 5501
108. Robin MB, Day P (1967) Adv Inorg Chem Radiochem 10: 247
109. Drüeke S, Chaudhuri P, Pohl K, Wieghardt K, Ding XQ, Bill E, Sawaryn A, Trautwein AX, Winkler H, Gurman SJ (1989) J Chem Soc Chem Commun 59
110. Ding XQ, Bominaar EL, Bill E, Winkler H, Trautwein AX, Drüeke S, Chaudhuri P, Wieghardt K (1990) J Chem Phys 92: 178
111. Gütlich P, Link R, Trautwein AX (1978) Mössbauer spectroscopy in transition metal chemistry Springer, Berlin Heidelberg New York
112. Papaefthymiou V, Girerd JJ, Moura I, Moura JJG, Münck E (1987) J Am Chem Soc 109: 4703
113. Mullinger RN, Cammack R, Rao KK, Hall DO, Dickson DPE, Johnson CE, Rush JD, Simopoulos A (1975) Biochem J 151: 75
114. Middleton P, Dickson DPE, Johnson CE, Rush JD (1978) Biochem J 88: 135
115. Moura JJG, Moura I, Kent TA, Lipscomb JD, Huynh BH, Le Gall J, Xavier AV, Münck E (1982) J Biol Chem 257: 6259
116. Moura I, Moura JJG, Münck E, Papaefthymiou V, Le Gall J (1986) J Am Chem Soc 108: 349
117. Surerus KK, Münck E, Moura I, Moura JJG, Le Gall J (1987) J Am Chem Soc 109: 3805
118. Münck E, Papaefthymiou V, Surerus KK, Girerd JJ (1988) In: Que L jr (ed) Metal clusters in proteins ACS Symposium Series No 372 Chap 15
119. Thomson AJ, Robinson AE, Johnson MK, Moura JJG, Moura I, Xavier AV, Le Gall J (1981) Biochim Biophys Acta 670: 93
120. Aizman A, Case DA (1982) J Am Chem Soc 104: 3269
121. Laskowski EJ, Frankel RB, Gillum WO, Papaefthymiou GC, Renand J, Ibers JA, Holm RH (1978) J Am Chem Soc 100: 5322
122. Cammack R, Dickson DPE, Johnson CE (1977) In: Lovenberg W (ed) Iron-sulphur proteins Academic New York, vol 3 p 283
123. Moulis JM, Auric P, Gaillard J, Meyer J (1984) J Biol Chem 259: 11396
124. Auric P, Gaillard J, Meyer J, Moulis JM (1987) Biochem J 242: 525
125. Lindahl PA, Day EP, Kent TA, Orme-Johnson WH, Münck E (1985) J Biol Chem 260: 11160
126. Vollmer SJ, Switzer RL, Debrunner PG (1983) J Biol Chem 258: 14284
127. Carney MJ, Papaefthymiou GC, Whitener MA, Spartalian K, Frankel RB, Holm RH (1988) Inorg Chem 27: 346
128. Carney MJ, Holm RH, Papaefthymiou GC, Frankel RB (1986) J Am Chem Soc 108: 3519
129. Middleton P, Dickson DPE, Johnson CE, Rush JD (1980) Eur J Biochem 104: 289
130. Noodleman L (1988) Inorg Chem 27: 3677
131. Banci L, Bertini I, Briganti F, Luchinat C, Scozzafava A, Oliver MV, Inorg Chim Acta (in press)
132. Rusnak FM, Adams MWW, Mortenson LE, Münck E (1987) J Biol Chem 262: 38
133. Bell SH, Dickson DPE, Rieder R, Cammack R, Patil DS, Hall DO, Rao KK (1984) Eur J Biochem 145: 645
134. Christner JA, Münck E, Kent TA, Janick PA, Salerno JC, Siegel LM (1984) J Am Chem Soc 106: 6786
135. Griffith JS (1972) Struct Bonding (Berlin) 10: 87
136. Girerd JJ, Papaefthymiou GC, Watson AD, Gamp E, Hagen KS, Edelstein N, Frankel RB, Holm RH (1984) J Am Chem Soc 106: 5941

137. Noodleman L, Case DA, Aizman A (1988) J Am Chem Soc 110: 1001
138. Kent TA, Huynh BH, Münck E (1980) Proc Natl Acad Sci 77: 6574
139. Darriet J, Mur J, Sanchez JP (1987) Nouv J Chimie 11: 21
140. Cannon RD, White RP (1988) Prog Inorg Chem 36: 195
141. Tweedle MF, Wilson LJ, Garzia-Iniguez L, Babcock GT, Palmer G (1978) J Biol Chem 253: 8065
142. Morgenstern-Badarau I, Laroque D, Bill E, Winkler H, Trautwein AX, Robert F, Jeannin Y (in press) Inorg Chem.
143. You JF, Snyder BS, Papaefthymiou GC, Holm RH (1990) J Am Chem Soc 112: 1067
144. Saak W, Henkel G, Pohl S (1984) Angew Chem, Int Ed Engl 23: 150
145. Kanatzidis MG, Salifoglou A, Coucouvanis D (1986) Inorg Chem 25: 2460
146. Noda I, Snyder BS, Holm RH (1986) Inorg Chem 25: 3851
147. Pohl S, Saak W (1984) Angew Chem, Int Ed Engl 23: 907
148. Kanatzidis MG, Hagen WR, Dunham WR, Lester RK, Concouvanis D (1985) J Am Chem Soc 107: 953
149. Murch BP, Boyle PD, Que L jr (1985) J Am Chem Soc 107: 6728
150. Jameson DL, Xie CL, Hendrickson DN, Potenza JA, Schugar HJ (1987) J Am Chem Soc 109: 740
151. Lippard SJ (1988) Angew·Chem., Int Ed Engl 27: 353
152. Drüeke S, Wieghardt K, Nuber B, Weiss J, Bominaar EL, Sawaryn A, Winkler H, Trautwein AX (1989) Inorg Chem 28: 4477
153. Sarel S, Avramovic-Grisaru S, Bauminger ER, Felner I, Nowik I, Williams RJP, Hughes NP (1989) Inorg Chem 28: 4187
154. Clegg GA, Fitton JE, Harrison PM, Treffry A (1988) Prog Biophys Mol Biol 36: 53
155. Crichton RR (1973) Angew Chem, Int Ed Engl 12: 57
156. Theil EC (1987) Annu Rev Biochem 56: 289
157. Dickson DPE, Reid NMK, Mann S, Wade VJ, Ward RJ, Peters TJ (1989) Hyperf Interactions 45: 225
158. Weir MP, Peters TJ, Gibson JF (1985) Biochim Biophys Acta 828: 298
159. Bauminger ER, Ofer S, Gedalia I, Horowitz G, Mager I (1985) Calcif Tissue Int 37: 386
160. St Pierre TG, Mann S, Webb J, Dickson DPE, Runham NW, Williams RJP (1986) Proc R Soc London B 228: 31
161. Ofer S, Nowik I, Bauminger ER, Papaefthymiou GC, Frankel RB, Blakemore RP (1984) Biophys J 46: 57
162. Bauminger ER, Nowik I (1989) Hyperf Interactions 50: 484
163. Bell SH, Weir MP, Dickson DPE, Gibson JF, Sharp GA, Peters TJ (1984) Biochim Biophys Acta 787: 227
164. Morup S, Dumesic JA, Topsoe H (1980) In: Cohen RC (ed) Application of Mössbauer spectroscopy II, Academic Press, New York, p 1
165. de Biasi RS, Devezas TC (1978) J Appl Phys 49: 2466
166. Ziethen HM, Winkler H, Schiller A, Schünemann V, Trautwein AX, Quazi A, Schmidt F (in press) Catalysis Today
167. Bill E, Haas C, Ding XQ, Maret W, Winkler H, Trautwein AX, Zeppezauer M (1989) Eur J Biochem 180: 111
168. Hewson WD, Hager LP (1979) In: Dolphin D (ed) The porphyrins, Academic, New York, vol 7, p 295
169. Dunford HB, Araiso T, Job D, Ricard J, Rutter R, Hager LP, Wever R, Kast WM, Boelens R, Ellfolk N, Ronnberg M (1982) In: Dunford HB, Dolphin D, Raymond KN, Shier L (eds) The Biological chemistry of iron. D Reidel, Dodrecht, p 337
170. Guengerich FP, MacDonald TL (1984) Acc Chem Res 17: 9
171. de Montelano O (1986) (ed) in: Cytochrome P450: Structure, mechanism and biochemistry Plenum, New York
172. Groves JT, Haushalter RC, Nakamura M, Nemo TE, Evans BJ (1981) J Am Chem Soc 103: 2884
173. Groves JT, Watanabe Y (1986) J Am Chem Soc. 108: 507
174. Traylor TG, Miksztal AR (1987) J Am Chem Soc 109: 2770
175. Boso B, Lang G, McMurry TJ, Groves JT (1983) J Chem Phys 79: 1122
176. Bill E, Ding XQ, Bominaar EL, Trautwein AX, Winkler H, Mandon D, Weiss R, Gold A, Jayaraj K, Hatfield WF, Kirk ML (1990) Eur J Biochem 188: 665

177. Buluggiu E, Vera A (1976) Z Naturforsch 31a: 911
178. Penner-Hahn JE, Eble KS, McMurry TJ, Renner M, Balch AL, Groves JT, Dawson JH, Hodges KO (1986) J Am Chem Soc 108: 7819
179. Schulz CE, Devaney PW, Winkler H, Debrunner PG, Doan N, Chiang R, Rutter R, Hager LP (1979) FEBS Lett 103: 102
180. Schulz CE (1989) private communication
181. Spaulding LD, Eller PG, Bertrand JA, Felton RH (1974) J Am Chem Soc 96: 982
182. Scholz WF, Reed CA, Lee YJ, Scheidt WR, Lang G (1982) J Am Chem Soc 104: 6791
183. Barkigia KM, Spaulding LD, Fajer J (1983) Inorg Chem 22: 349
184. Gans P, Buisson G, Duée E, Marchon JC, Erler BS, Scholz WF, Reed CA (1986) J Am Chem Soc 108: 1223
185. Czernuszewicz RS, Macor KA, Li XY, Kincaid JR, Spiro TG (1989) J Am Chem Soc 111: 3860
186. Goodenough JB (1963) In: Magnetism and the chemical bond, Wiley, New York

Tumor-Inhibiting Bis(β-Diketonato) Metal Complexes. Budotitane, cis-Diethoxybis(1-phenylbutane-1,3-dionato)titanium(IV)

The First Transition Metal Complex After Platinum to Qualify for Clinical Trials

B.K. Keppler, C. Friesen, H.G. Moritz, H. Vongerichten, E. Vogel

Anorganisch-Chemisches Institut der Universität Heidelberg, Im Neuenheimer Feld 270, 6900 Heidelberg, FRG

Cisplatin, cis-diamminedichloroplatinum(II), and carboplatin, cis-diammine(cyclobutane-1,1-dicarboxylato)platinum(II), are the first drugs from inorganic chemistry to have come under routine clinical use in medical oncology. Their antitumor activity ranges from testicular carcinomas, ovarian carcinomas, and tumors of the head and neck to bladder tumors. However, the spectrum of indication is fairly limited. There is no or only insufficient antitumor activity in tumors which account for the major share of cancer mortality today, e.g. lung tumors and gastrointestinal tumors. Direct derivatives of cisplatin such as carboplatin have only led to a limited reduction or change in drug toxicity. In most cases, the toxicity pattern has changed from nephrotoxicity to myelotoxicity. New metal complexes are now being developed which are designed to supplement the spectrum of indication of platinum complexes. Among non-platinum complexes, budotitane (INN), cis-diethoxybis(1-phenylbutane-1,3-dionato)titanium(IV), is among the most advanced. It is undergoing clinical trials today. Extensive investigations into structure–activity relations have clearly shown a dependence of the activity on the central metal and the diketonato ligand. The tumor-inhibiting effect decreases in the order titanium > zirconium > hafnium > molybdenum > tin > germanium. Antitumor activity is also highly dependent on the nature of the diketonate used. Ligands substituted with planar aromatic ring systems such as the phenyl groups in budotitane are advantageous. Most of the tumor-inhibiting bis(β-diketonato) complexes are cis-configurated. The cis-configurated compounds with an unsymmetrically substituted β-diketonate as ligand are in an equilibrium between three possible cis-isomers in solution at room temperature, due to the fact that the diketonate can rotate via a twist mechanism. The easily hydrolizable group in these complexes does not play a major role in antitumor activity, but it is important for the galenic formulation in the clinic. The ethoxy group as leaving group in budotitane hydrolizes at a slower rate than the corresponding halides.

The best antitumor effects could be obtained with titanium and a diketonato ligand substituted with phenyl groups. Budotitane is highly active in several transplantable tumors and shows promising effects in an autochthonous colorectal tumor model, which is highly predictive for the clinical situation. Side-effects include mild hepatotoxicity and nephrotoxicity. These findings have been confirmed in clinical phase I studies. A phase II study is now in preparation. If preclinical antitumor activity in colorectal tumors can be confirmed in the clinic, this would lead a considerable step forward in the chemotherapy of cancer.

List of Abbreviations

acac   = acetylacetonato = pentane-2,4-dionato
bzac   = benzoylacetonato = 1-phenylbutane-1,3-dionato
dik   = diketonate
Et   = C_2H_5
Hal   = halogen
i.p.   = intraperitoneally
i.v.   = intravenously
ICR   = trans-Imidazolium-[tetrachlorobisimidazoleruthenate(III)]
         $HIm(RuIm_2Cl_4)$
INN   = International Nonproprietary Name
LD_{50} = Dose at which 50% of animals will die
s.c.   = subcutaneously

1 Introduction

Since early history, metal complexes have been used as pharmaceutical agents. However, therapeutic efficacy in today's meaning of the term was not confirmed until the developement of salvarsan (1910), used in cases of syphilis, and some organic mercury compounds such as novasurol (1919) and salyrgan (1924), used as diuretic agents. These drugs have gradually been replaced by organic compounds with better activity. Today, the most important drugs from the field of inorganic chemistry include auranofin (INN), (2,3,4,6-tetra-*O*-acetyl-1-thio-1-β-

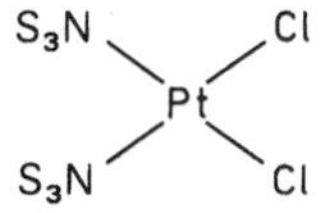

$$S_3N\diagdown \quad \diagup Cl$$
$$Pt$$
$$S_3N\diagup \quad \diagdown Cl$$

Cisplatin (INN)

Fig. 1. Cisplatin (INN), *cis*-Diamminedichloroplatinum(II)   *cis*-Diamminedichloroplatinum (Ⅱ)

D-glucopyranosato)(triethylphosphine)gold(I), active against primary chronic polyarthritis (PCP) [1, 2], sodium nitroprusside, nipruss, disodium pentacyano-nitrosylferrate(II)dihydrate, $Na_2[Fe(NO)(CN)_5]xH_2O$, used as an emergency drug in cases of hypertensive crises, lithium salts, used in psychiatry, bismuth salts, used in cases of infections with *Heliobacter pylori*, magnesium salts, used in cardiology, many preparations for local application in dermatology and gastroenterology, and various metal salts for the prevention of deficiencies. In the treatment of cancer, only one drug from inorganic chemistry is used on a routine clinical basis – cisplatin (INN), *cis*-diamminedichloroplatinum(II) (Fig. 1). This drug was synthesized for the first time by Michele Peyrone and was published in 1844 in the "Annals of Chemistry and Pharmacy" [3].

In 1969, Barnett Rosenberg discovered the tumor-inhibiting properties of cisplatin. While examining the influence of an electrical field on bacteria growth, Rosenberg found filamentous growth of *Escherichia coli* bacteria, which was caused by the experimental conditions that favoured the formation of *cis*-configurated platinum complexes. Observing a selective influence on cell division along with an unrestrained growth, Rosenberg concluded that these compounds might also be capable of inhibiting tumor growth [4, 5]. Cisplatin soon qualified for clinical studies and showed good effects on a number of different tumors. Today, cisplatin is mainly used in combination with other anticancer agents, and its spectrum of indication includes testicular carcinomas, ovarian carcinomas, bladder tumors, and tumors of the head and neck. Testicular carcinomas had almost invariably been fatal before cisplatin was discovered. Today, most cases are curable, thanks to cisplatin.

Unfortunately, the high activity of cisplatin is limited to tumors which are relatively rare. The tumors that are very common and that account for the major share of cancer mortality today, e.g. lung tumors and adenotumors of the gastrointestinal tract, are not affected by cisplatin, or only to a negligible extent. However, the example of cisplatin shows that it is possible to find new drugs – in this case from the field of inorganic chemistry – that are capable of curing specific types of tumors. This has been a considerable impetus for the inorganic chemist to search for new metal complexes with good antitumor activity, which may be used to manage the most common tumors.

The basic strategies in the development of new tumor-inhibiting metal complexes may be summed up as follows:

– *synthesis and activity screen of "direct" cisplatin derivatives*

– *linking of tumor-inhibiting platinum compounds or of other tumor-inhibiting metal complexes with carrier systems to achieve accumulation in particular tissues.*
– *synthesis of new metal complexes with different central metals*

The first of these three strategies is not particularly promising, because owing to a similar mechanism of action, the therapeutic efficacy of direct derivatives of cisplatin is unlikely to differ substantially from that of the parent compound. Certainly the synthesis of cisplatin derivatives is justified when trying to decrease toxicity or increase selectivity. The second strategy has produced hormone-linked platinum derivatives, which have an affinity to hormone receptor-positive tumors, and osteotropic platinum compounds [6, 7].

New tumor-inhibiting metal complexes with different central metals, that is, non-platinum compounds, may be expected to act against tumors other than those affected by cisplatin, because they have different chemical properties. The problem is that there are no established structure–activity relationships for metals other than platinum, and thus it is much more difficult in this field to find compounds that exhibit any tumor-inhibiting properties at all. However, this third strategy has meanwhile produced some success. Ruthenium, tin, gold, and titanium compounds must be mentioned here [8–17]. In the field of titanium chemistry, Budotitane (INN) is the most advanced anticancer agent. It is already under clinical trials [17, 18]. Other research groups have investigated organometallic titanium compounds [19].

Budotitane (Fig. 2) belongs to the class of bis(β-diketonato) metal complexes. The antitumor activity of some representatives of these complexes was reported on as early as 1982. We described the *cis*-dihalogenobis(1-phenyl-1,3-butanedionato)titanium(IV) complexes with fluoride, chloride, and bromide as further ligands. These complexes effected a doubling and even a trebling of survival time of treated animals in the Walker 256 carcinosarcoma and in a transplantable murine leukemia [20, 21].

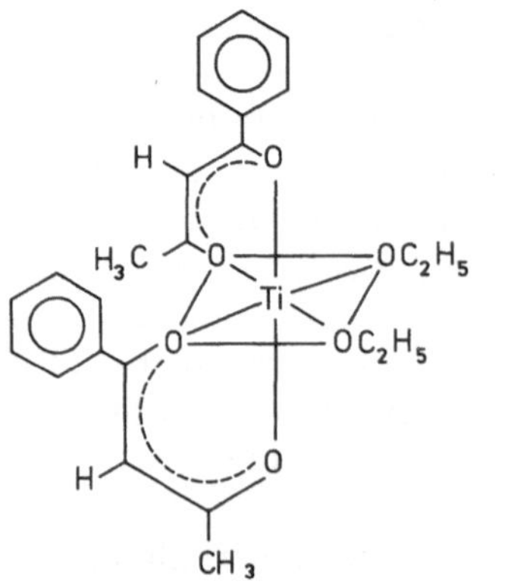

Budotitane (INN)

cis-Diethoxybis(1-phenylbutane–1,3–dionato)titanium(**IV**)

Fig. 2. Budotitane (INN), *cis*-Diethoxybis(1-phenylbutane-1,3-dionato)titanium(IV), Ti(bzac)$_2$(OEt)$_2$

In later studies budotitane, *cis*-diethoxybis(1-phenylbutane-1,3-dionato)titanium(IV), Ti(bzac)$_2$(OEt)$_2$, was selected from about 200 representative compounds of the class of bis(β-diketonato) metal complexes for further development and, finally, clinical studies. In the following we will describe the most important stages in the development of this drug [17 18, 22–27].

2 Synthesis and Characterization of Budotitane and Bis(β-diketonato) Metal Complexes

The bis(β-diketonato) metal complexes can be synthesized from the corresponding metal tetrahalogenides and the diketonates in an anhydrous organic solvent, as presented in Fig. 3. An exception is the corresponding molybdenum compounds, where the basis is molybdenum pentachloride. The compounds synthesized this way are six-coordinated with quasi-octahedral configuration, which may come either in the *cis*- or the *trans*- form (Fig. 4). The many metal complexes synthesized were characterized by the methods described below.

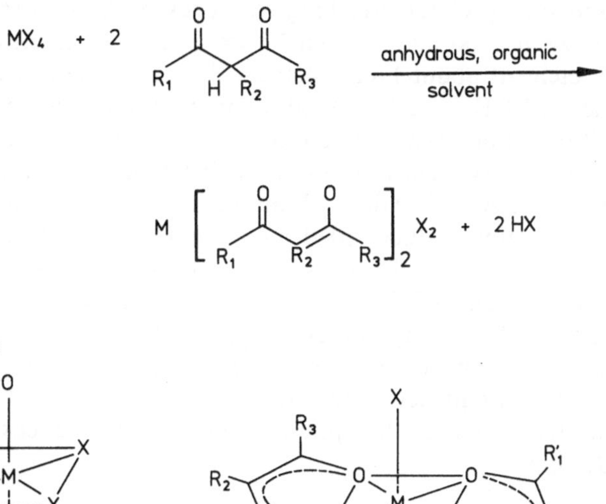

Fig. 3. General synthesis of M(β-diketonato)$_2$X$_2$ complexes. X = Hal or OR; M = Ti, Zr, Hf, Ge, Sn; R = organic group or H

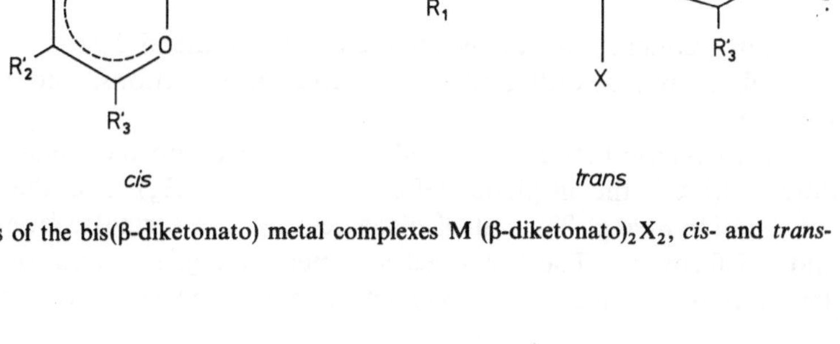

cis   trans

Fig. 4. Structures of the bis(β-diketonato) metal complexes M (β-diketonato)$_2$X$_2$, *cis*- and *trans*-configuration

3 Spectroscopic Characterization

3.1 IR Spectroscopic Characterization

IR Region between 650 and 4000 cm^{-1}

β-Diketones are found in a mixture of keto and enol forms, with the equilibrium being markedly shifted towards the enol form [28].

In the case of acetylacetone, the C–O vibrational frequency of the keto form is found around 1709 cm^{-1} and that of the enol form at around 1613 cm^{-1}. In the case of benzoylacetone and dibenzoylmethane it is around 1600 cm^{-1} and around 1592 cm^{-1}. In the titanium chelate, however, these C–O bands are shifted by about 50–100 cm^{-1} towards lower frequencies (1550–1590 cm^{-1}) in comparison to the enol form of the free ligand. The absorption band that is characteristic of the free ligand disappears, which proves that the complex is six-coordinated, with a bidentate ligand.

In the case of M(acac)$_2$X$_2$ (M = Ti, Ge, Sn, Zr, Hf; X = Hal, OR), the symmetric stretching vibration splits into a doublet (approx. 1575 cm^{-1} and approx. 1554 cm^{-1}), which is a result of the coupling between the metal atom and the ν_s (C–O) stretching frequency of the two chelate rings. In the region between 1500 and 1600 cm^{-1} we find a less intense band of an asymmetric stretch. At about 1408–1460 cm^{-1} there is the degenerate deformation vibration of CH$_3$, and at 1340–1370 cm^{-1} there is the symmetric deformation vibration of CH$_3$. Furthermore, in the region between 1310 and 1370 cm^{-1} we find the asymmetric C–O stretching vibration, at 1250–1310 cm^{-1} the symmetric C–C stretching vibration, at 1160–1210 cm^{-1} the δ (C–H) in-plane deformation, at 1000–1040 cm^{-1} ρ_r(CH$_3$) rocking, at 930–980 cm^{-1} the stretch of ν (C–CH$_3$) and ν (C–C), and at 740–820 cm^{-1} there is the out-of-plane deformation π (C–H). Overlappings are frequently seen. When we compare the spectra of the complexes of the type M(dik)$_2$X$_2$, M = Ti, Ge, Zr, Hf, Sn, with the same diketone and a different ligand X, we find an agreement of the spectra in the high-frequency region. In the case that X is not halogen but OR, there is another intense absorption band, e.g. in the case of Ti(bzac)$_2$(OEt)$_2$, at 910 cm^{-1}.

IR Region between 650 and 185 cm^{-1}

The low-frequency region is characteristic of metal–ligand vibrations, that is, chelate ring deformations, metal–oxygen vibrations, and metal–halogen vibrations.

In the region between 640 and 700 cm^{-1} we find the metal–oxygen vibrations ν(M–O), the in-plane deformation δ (C–CH$_3$), and the in-plane ring deformation δ(ring). The out-of-plane deformation π(ring) is found between 540 and 580 cm^{-1}. The symmetric metal–oxygen vibration occurs at 450–470 cm^{-1}, and the asymmetric metal–oxygen vibration occurs at

$430–450\ cm^{-1}$. However, we do not always find separate bands, as the vibrations overlap in most cases. Another M–O vibration is found in the region between 330 and $370\ cm^{-1}$. The metal–halogen vibrations occur in different regions: in the case of fluorine complexes, an intense band is found between 605 and $650\ cm^{-1}$. A comparable band does not exist for the chlorine complexes. The metal–chlorine vibration is found only in the region between 330 and $390\ cm^{-1}$. This band is very intense, because the metal–chlorine vibration overlaps with the metal–oxygen vibration. Finally, the metal–bromine vibration is found in the region between 305 and $340\ cm^{-1}$. The overlapping of the bands makes it impossible in most cases to distinguish between *cis-* and *trans-* by the number of Ti-Hal bands. The IR spectroscopic findings for those compounds which are already known from the literature are in agreement with the IR findings reported there [29–34].

3.2 *^{1}H-NMR* Spectroscopic Characterization

Anisotropic Effects of the Diketonato Chelate Ring

In the six-coordinated $Ti(dik)_2X_2$ complexes, the 1,3-diketonates form a six-membered, planar chelate ring with the metal. There has been controversial discussion about whether an enol character must be ascribed to the chelate ring [35, 36] or whether it must be considered as a benzoid system, with the transition metal atom participating in the aromatic system by forming π bonds [37–41]. It is possible to carry out reactions at the metal chelate ring that otherwise are observed only with aromatic molecules [37, 42–49].

The planarity or "near" planarity of the ring can be proved by X-ray analyses [50]. ^{1}H-NMR measurements show downfield shifts of the CH_3- and the ring proton (–CH=) of the metal chelate ring, in contrast to the free 1,3-diketone, which is caused by the anisotropic effect of the chelate ring. This is evidence of a certain aromaticity of the system.

The methino proton of the free benzoylacetone, for example, shows a chemical shift of 6.2 ppm, whereas it is 6.3 ppm in $Ti(bzac)_2(OEt)_2$. When F is substituted for OEt, one finds 6.5 ppm, in the case of Cl 6.68 ppm, and in the case of Br 6.7 ppm [17].

4 Isomer Equilibria of the $M(\beta\text{-diketonato})_2X_2$ Complexes in Solution

The six-coordinated, "octahedrally" configured compounds can have both the *cis-* and *trans-*configuration (Fig. 4). The *cis-*conformers are chiral. It is surprising but the *cis-*configuration is usually energetically favoured, even though the *trans*-isomer should really be favoured for steric reasons. The greater stability of the *cis*-configuration should be attributed to the π back donation (p_π-X – d_ε-Ti), because in the case of the *cis*-configuration all three d_ε orbitals can be included in bonds by the titanium, whereas in the case of the *trans*-configuration, only two of these orbitals can be occupied [51].

The number of possible isomers in the *cis*- and the *trans*- form depends on whether the bound diketone in 1- and 5-position has the same or different substituents. For the sake of brevity, similarly substituted diketones – such as acac – are termed "symmetric" β-diketones, and dissimilarly substituted "asymmetric" diketonates. The *cis*-isomer with symmetric diketonate groups has C_2 symmetry. The molecule shows chirality. The complex exists in two enantiomeric forms. The *trans*-isomer, however, belongs to the symmetry group D_{2h} and thus is achiral.

In accordance with this, the ^{1}H-NMR spectra of symmetrically substituted diketonates, such as the acetylacetonates, should show for the *cis*-isomer two signals of the same intensity for the non-equivalent CH_3 groups and one signal for the ring methino proton. In contrast to this, the *trans*-isomer should show only one signal for the methino group and one for the methyl groups. However, the ^{1}H-NMR spectra at room temperature show only one signal in all cases. During cooling, the signal for the non-equivalent CH_3 groups splits up into two signals. In addition, there is one signal for the methino ring proton. That is because the majority of these complexes are *cis*-configurated. On account of the molecular fluctuation, which probably occurs via a so-called twist mechanism with a trigonal bipyramid as transition state, the CH_3 groups of the diketonate ligands exchange their positions, with the result that different signals for the two methyl groups can only be recognized at a low temperature, when this fluctuation is sufficiently "frozen". The coalescence temperature serves as a basis for calculating the free activation enthalpy of the conformational change, which is about 11 to 12 kcal/mol. The coalescence temperature is markedly dependent on the central metal used. Thus, in the case of $Sn(acac)_2Cl_2$, for example, we find the split CH_3 signal in the favoured *cis*-form even at room temperature [52].

Conditions are much more interesting in the case of the bis(β-diketonato) complexes with asymmetric substitution, such as the benzoylacetonato complexes, which are among those compounds within this class of substances which have the best antitumor activity. In principle, there are three isomers in the *cis*-form and two in the *trans*-form (Fig. 5).

The three *cis*-isomers each form three enantiomeric pairs, which are diastereomers to each other. In contrast to this, the two *trans*-isomers are achiral. We expect the *cis*-form to show four ^{1}H-NMR signals for the different methyl groups. These are indeed observed in low-temperature experiments. The same applies also to the methino ring protons, owing to their different environment with phenyl and CH_3 groups (Fig. 6). Also, the splitting of the quadruplet of the TiOEt group of budotitane shows a characteristic temperature dependence, which is due to the chemical non-equivalence of the two diastereotopic $-CH_2$ protons, which are found in vicinity to a chiral centre (Fig. 7).

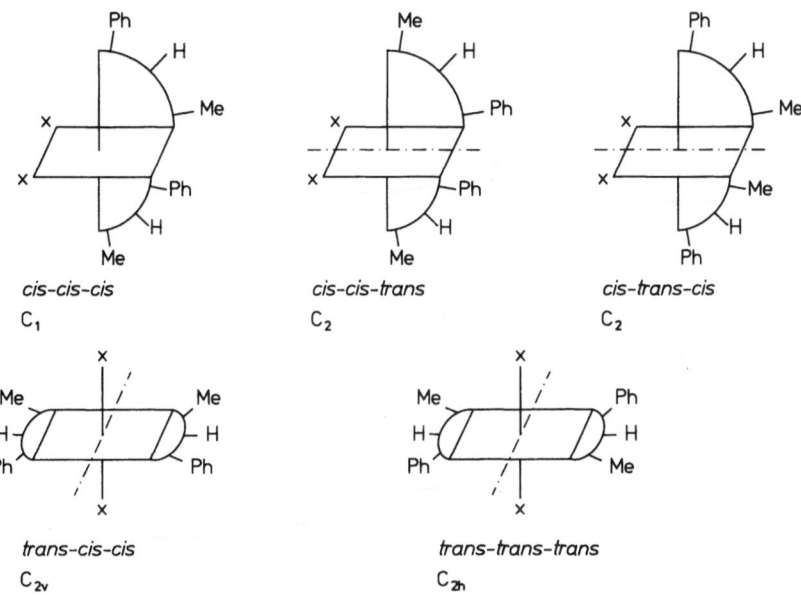

Fig. 5. Number of isomers within the *cis-* and the *trans-* form of asymmetrically substituted bis(β-diketonato) metal complexes. This example: a benzoyl acetonato complex

Table 1 shows that of all the metal (β-diketonates) listed, which have benzoylacetone or its derivatives as ligand, only those with iodide and *p*-dimethylaminophenoxy as hydrolizable group have the *trans-* form. All other complexes are *cis*-configurated. In the cases of comparable diketonato ligands the coalescence temperature increases, depending on the hydrolizable group X, in the order I < Cl < F < OR. If we compare the coalescence temperatures in complexes with the same diketonato ligand and different metals, we find signals that are split already at room temperature in the cases of Zr, Hf, Mo, Ge and Sn. This is evidence that these complexes are kinetically more stable.

5 Hydrolysis of Bis(β-diketonato) Metal Complexes

Hydrolysis of the bis(β-diketonato) metal complexes is especially important in terms of storage and galenic[1] formulation for application in the clinic. Every compound which shall be applied intravenously must be transferred into a watery medium. Other methods, e.g. oral application and intramuscular injection, cannot be carried out with the majority of tumor-inhibiting drugs, and

---

[1] The galenic formulation of a drug means the processing of plain chemical substances to make them applicable in the clinic

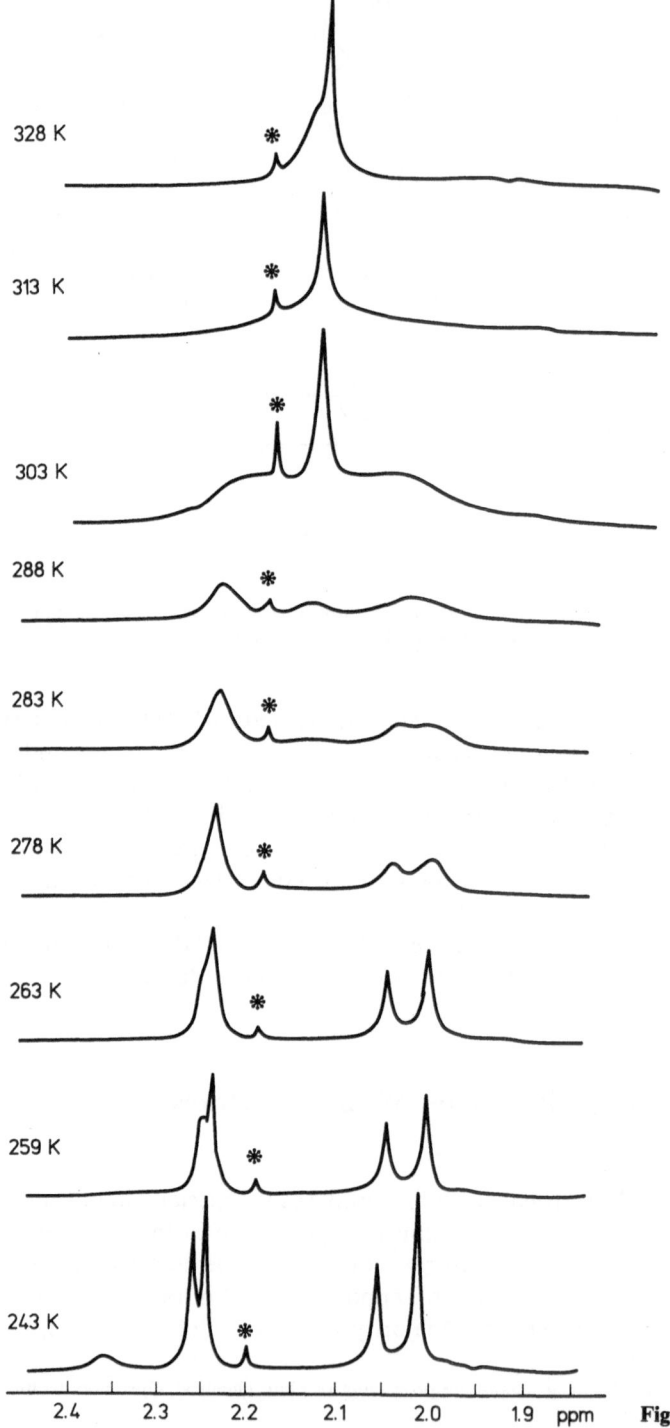

Fig. 6a.

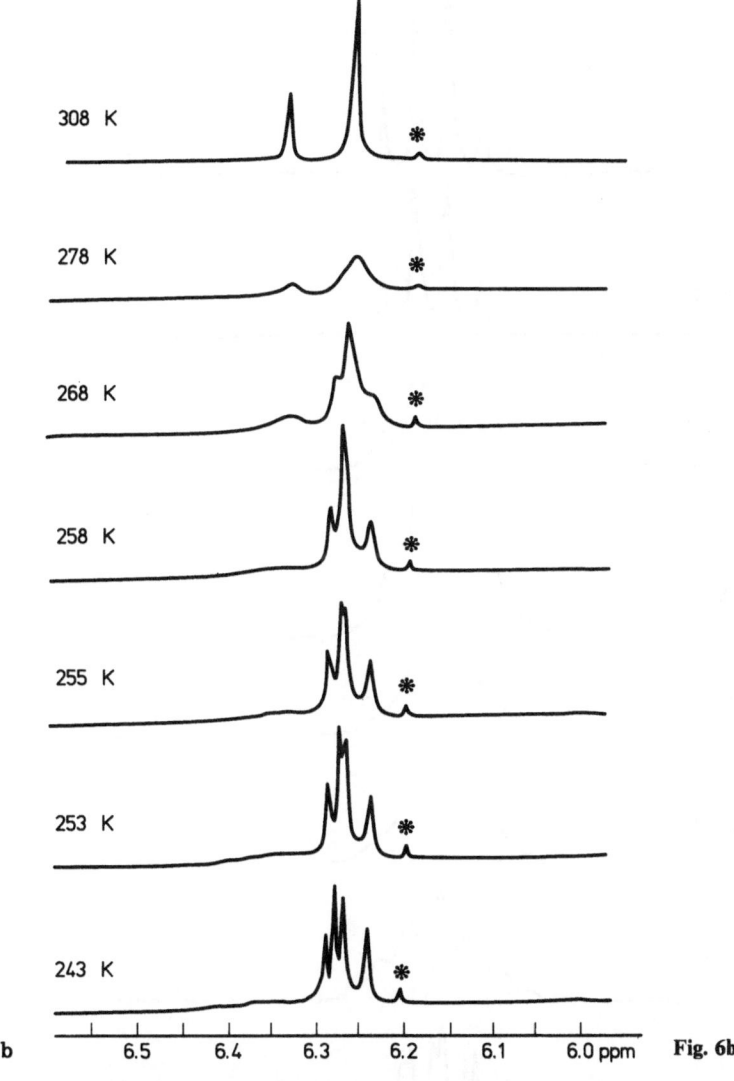

308 K

278 K

268 K

258 K

255 K

253 K

243 K

b   6.5   6.4   6.3   6.2   6.1   6.0 ppm   **Fig. 6b.**

◀ **Fig. 6a, b.** Temperature dependence of the 200 MHz ^{1}H-NMR spectrum of budotitane in CDCl$_3$ showing the coalescence of the peaks for methyl groups (**a**) and non-equivalent methino groups (**b**). (a)* = signal of the methyl group of the free ligand. (b)* = signal of the methino proton of the free ligand

neither can they be carried out with the bis(β-diketonato) metal complexes. The intramuscular application of substances that are difficult to dissolve in water, in the form of lipid suspensions, is often associated with marked variations in absorption. This would be too great an element of uncertainty in the administration of tumor-inhibiting drugs, because dosages must often be on the threshold

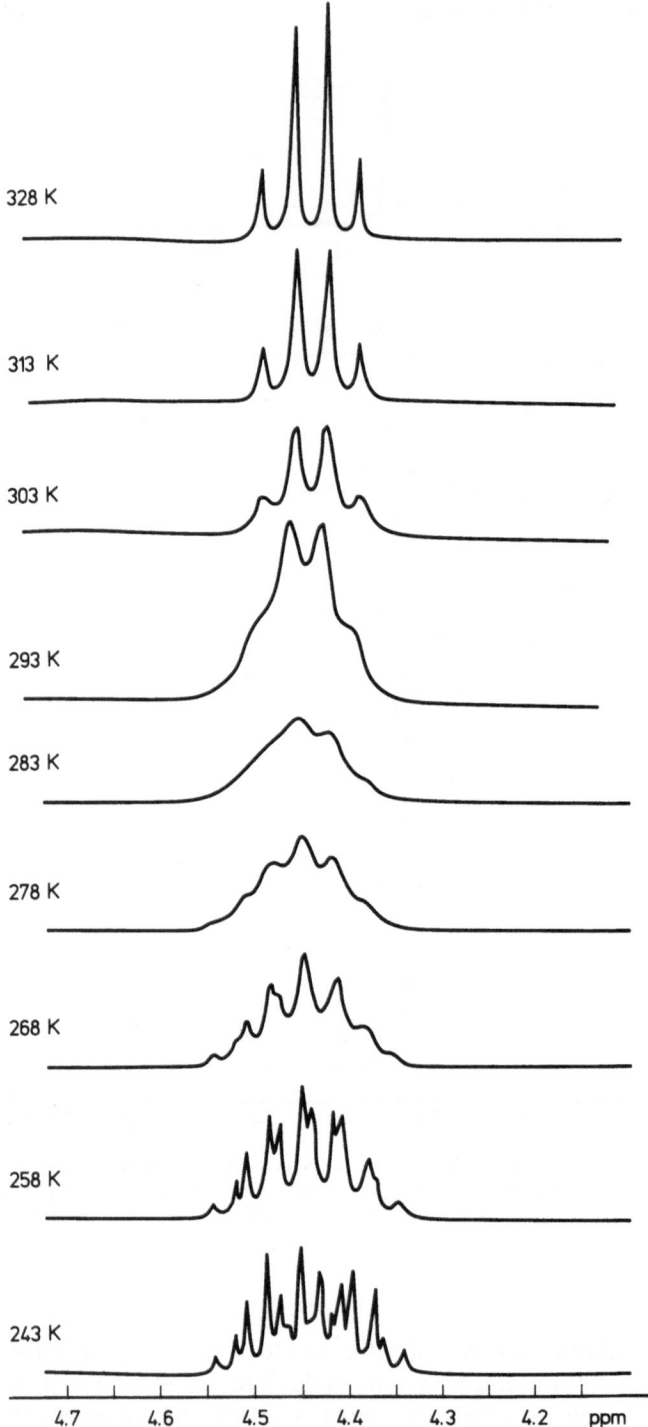

Fig. 7. The splitting of the quadruplet of the TiOEt group of budotitane shows a characteristic temperature dependence, which is due to the chemical non-equivalence of the two diastereotopic-CH_2 protons, which are found in vicinity to a chiral centre

Table 1. Chemical shift and splitting of the signals of the methino and methyl protons of the β-diketonato ligand in complexes of the type $M(dik)_2X_2$ [17]. – CH = signal of the methino ring proton (ppm); –CH₃ signal of the methyl protons (ppm); * Number of signal splittings below the coalescence temperature; ** Width of splitting (ppm); T = coalescence temperature in °C (for CH₃). Solvent: CD_2Cl_2

M	dik	X	–CH =	*	**	–CH₃	*	**	T	MHz
Ti	bzac	OEt	6.30	4	0.25	2.15	4	0.25	19	200
Ti	bzac	Cl	6.68	4	0.20	2.29	4	0.25	−24	200
Ti	bzac	I	6.80	2	0.27	2.31	2	0.13	−65	300
Ti	bzac	O–⟨C₆H₄⟩–NMe₂ · HCl	6.40	2	0.41	2.05	2	0.25	20	300
Ti	bzac	O–⟨cyclohexenyl-H⟩	6.83	4	0.10	2.15	4	0.35	24	200
Ti	bzac	O–⟨methylenedioxyphenyl⟩	6.32	4	0.08	2.15	4	0.23	18	200
Zr	bzac	Cl	6.40	4	0.40	2.26	4	0.25	> RT	90
Hf	bzac	Cl	6.45	4	0.49	2.26	4	0.25	> RT	90
Mo	bzac	Cl	6.32	—	—	2.31	4	0.35	> RT	90
Ge	bzac	Cl	6.45	—	—	2.26	4	0.31	> RT	90
Sn	bzac	Cl	6.40	—	—	2.29	4	0.35	> RT	90
Ti	(Ph–CO–CH=C(O)–C(CH₃)₃)	F	6.75	4	0.04	1.2	4	0.05	11	200
Ti	(Ph–CH₂–CO–CH=C(O)–)	OEt	5.43	4	0.30	1.91	4	0.39	25	200
Ti	(Ph–CO–C(Cl)=C(O)–CH₃)	Cl	—	—	—	2.40	4	0.30	−10	200
Ti	(thienyl–CO–CH=C(O)–CH₃)	Cl	6.53	4	0.05	2.28	4	0.24	−14	200
Ti	((CH₃)₃C–CO–CH=C(O)–CH₃)	OEt	5.72	4	0.15	2.00	4	0.22	> RT	200·

of the maximum tolerated dose in order to achieve sufficient therapy. In addition, intramuscular or subcutaneous injection is accompanied by marked local toxicity with tissue necrosis, at least in animal experiments. Oral application of bis(β-diketonato) metal complexes would lead to a partial hydrolysis in

$M(\text{diket})_2 X_2 \xrightarrow[\quad]{H_2O} \quad [M(H_2O)(\text{diket})_2 X]^+ X^-$

$\xrightarrow[\quad]{\quad} M(OH)(\text{diket})_2 X \quad + HX$

$\xrightarrow[\quad]{H_2O} \quad [M(OH)(H_2O)(\text{diket})_2]^+ X^-$

$\xrightarrow[\quad]{\quad} M(OH)_2(\text{diket})_2 \quad + HX$

$\xrightarrow[\quad]{-H_2O} \text{polymers} \quad\longrightarrow\quad ----\blacktriangleright-----\blacktriangleright MO_2$

X = OR, halogen   M = Ti, Zr, Hf, Ge, Sn

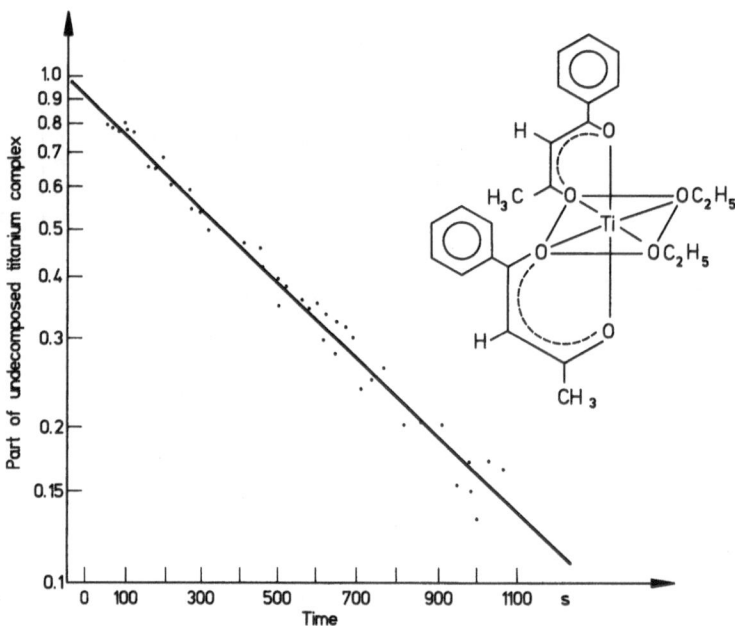

Fig. 8. Scheme of hydrolysis of $M(\beta\text{-diketonato})_2 X_2$ complexes. X = Hal, OR; M = Ti, Zr, Hf, Ge, Sn. *and* Pseudo first order decay of $Ti(bzac)_2(OC_2H_5)_2$ during hydrolysis of 20.05 mg of complex in 400 µl of CD_3OH + 20 µl of D_2O at 22 °C. The reaction was followed by measurement of the intensity of the NMR signal of the $Ti-O-CH_2-CH_3$ protons

the alkaline environment of the small intestine, and this would also make an exact dosage difficult. Animal experiments have confirmed this. Hence we can apply these metal complexes only intravenously, and in an aqueous medium.

The reaction of $M(\beta\text{-diketonato})_2 X_2$ complexes with water is often peculiar. For example, $Ti(bzac)_2(OEt)_2$ is insoluble in water, and it can be suspended in water completely undecomposed. This is probably caused by passivation by a monomolecular hydroxide film. If, however, it is dissolved in a water-soluble organic solvent such as acetonitrile, and if then water is added, the complex reacts under hydrolysis according to the scheme shown in Fig. 8. Substitution of the leaving group OEt occurs relatively quickly, as can be gathered from NMR investigations with budotitane (cf. Fig. 8) [53].

The reaction rate was measured by integration of the ^{1}H-NMR signals of the –CH$_2$-protons of the complex-bound ethoxy groups, which have a different chemical shift from the –CH$_2$-protons of the ethanol which is released during hydrolysis.

Decomposition was carried out in acetonitrile, adding defined amounts of water. When adding different amounts of water, different reaction rates of pseudo first order are obtained for the hydrolytic decomposition of the titanium complex in mixtures of acetonitrile and water. This is the basis for calculating the half life – in terms of hydrolysis of the ethoxy groups – of the titanium complex in pure water as $T_{1/2}$ = ca. 20 sec. Substitution of the diketo group under similar conditions, however, can be observed only at a later time, i.e. after about two hours and thirty minutes. Ti(bzac)$_2$(OEt)$_2$, cis-diethoxybis(1-phenylbutane-1,3-dionato)titanium(IV), budotitane, thus is markedly sensitive to hydrolysis. However, measurements comparing it to other diketonato complexes have proved that this substance is among the most stable complexes, and therefore it had to be given preference over the others in further development for galenic reasons.

When the hydroxy group is replaced by fluorine, chlorine, bromine, or iodine, hydrolysis becomes even more rapid. Replacement of the central metal titanium by zirconium, hafnium, germanium, tin, or molybdenum also results in accelerated hydrolysis. A broad variation of the diketonato ligands also has an adverse effect on hydrolysis. In order to examine this, we introduced phenyl groups or alkyl and cycloalkyl groups, which in turn were also substituted with different ligands. The substitution of the hydrolizable group, using bigger and bulkier leaving groups, did not lead to a decrease in hydrolysis either, not even when the leaving groups contained phenyl groups or tertiary butyl groups. The introduction of groups which were also meant to increase water-solubility, such as $-(OC_2H_4)_nOC_2H_5$, n = 1–3, $-OCH_2CH_2OH$, $-OCH_2CH_2COOH$, $-OCH_2CH_2SO_3H$, $-OCH_2CH_2N(Et)_2$, did not result in any improvements either.

Apparently, hydrolysis is the only reaction to play a role during storage. Thus we had to investigate the products of hydrolysis and possible measures to avoid hydrolysis.

6 Structural Characterization of the Products of Hydrolysis and Galenic Consequences

The primary product of the hydrolysis of two ethoxy groups of budotitane can be obtained in a pure form by reaction of the complex with water in the stoichiometric ratio of 1:2 in acetonitrile and subsequent evaporation of the solvent and recrystallization from CCl_4/petroleum ether. Its composition is $[Ti(bzac)_2 O]_n$, which is probably produced by oligomerization of the intermediate $Ti(bzac)_2(OH)_2$ formed on reaction of budotitane with water [53].

$$Ti(bzac)_2(OEt)_2 + 2H_2O \rightarrow Ti(bzac)_2(OH)_2 + 2C_2H_5OH$$

$$n\,Ti(bzac)_2(OH)_2 \rightarrow 1/n\,[Ti(bzac)_2O]_n + nH_2O$$

The same product is also obtained by thermolysis of budotitane when storing it openly in a drying oven at 100 °C. In this process, stoichiometric expectations would include either formation of diethyl ether or of ethanol/ethylene in the ratio of 1:1 [53].

The IR spectra of both the thermolysis and the hydrolysis products are identical. The C–H vibrational frequencies between 2800 cm^{-1} and 3000 cm^{-1} and the C–O vibrations at 1060 cm^{-1} and in the region from 1110 to 1140 cm^{-1} decrease markedly. Also, an intense absorption band at 910 cm^{-1} disappears with the hydrolysis of the ethoxy groups. Instead, a new, broad and intense band forms at 810 cm^{-1}. The vibration at 810 cm^{-1} could correspond to a Ti–O–Ti vibration. Determining molecular weight by vapour pressure osmometry in benzene gave a median molecular weight of 940 mass units for the hydrolysis product. Thus the degree of polymerization is n = 2.43, i.e. a mixture of cyclic dimers and trimers must be assumed. Since the crystal quality of the decomposition products has so far been insufficient for X-ray investigations, this could not be proved with absolute certainty, but structures analogous to the dimer, with acetylacetone as ligand, have been resolved by X-ray analyses and have been described in the literature [54]. These oligomers may be a good model for the decomposition products that might possibly form during the storage of budotitane. However, since our NMR investigations have shown that adding ethanol to these decomposition products causes a complete reverse reaction to the intact budotitane, the development of such hydrolysis products can easily be avoided by adding a low excess amount of ethanol to the galenic formulation of budotitane.

Having solved the problems of storage, we were now faced with the difficulty of dissolving the compound in water for application to the patient without hydrolysis. How we managed this will be described in the following.

7 Galenic Formulation of Budotitane

A drug which may be used in the clinic or in preclinical toxicological and pharmacological studies must be both water-soluble and relatively insensitive to hydrolysis, or one has to find a galenic formulation for the substance which guarantees these properties. In our case, the use of CremophorEL, a glycerinepolyethylene-glycolericinoleate (BASF), was successful. Also 1,2-propylenglycol was added, and thus water-soluble galenic preparations of the drug, so-called coprecipitates, were able to be produced [25]. Budotitane, Cremo-phorEL, and propylenglycol are dissolved separately in water-free ethanol in the weight ratio of 1:9:1. The solutions are then mixed and evaporated at 30–40 °C. The product is what is known as a coprecipitate. In this coprecipitate, budotitane is enveloped between layers of solubilizer. This can be dissolved in water under formation of micelles. The micelles take up the drug, protect it from hydrolysis, and make it water-soluble to a sufficient extent.

When filtering watery emulsions of the coprecipitate through a 5 nm mem-brane filter, which keeps the micelles quantitatively behind, the filtrate contains only a maximum amount of 0.7% of the titanium contained in the coprecipitate solution. The conclusion is that the titanium complex must be virtually quantit-atively fitted in the CremophorEL micelles. If the ultrafiltration experiment is carried out not with pure watery emulsions of the coprecipitate but with solutions that also contain human serum (1 mg of budotitane in 2 ml of serum), the filtrate will contain up to 10% of the amount of titanium used. Under these conditions, the complex is released from the micelles by the serum proteins [53].

In order to examine the decomposition of the complex in the watery micellar emulsion, UV spectra were taken and the levels of intact complex were calcu-lated by component analysis. Other methods, e.g. ^{1}H-NMR spectroscopy, cannot be used owing to the marked overlapping of the spectra of complex and solubilizer. Figure 9 shows that the micellar emulsion, after an initial rapid decomposition up to about 80% of intact titanium complex, is stable over many

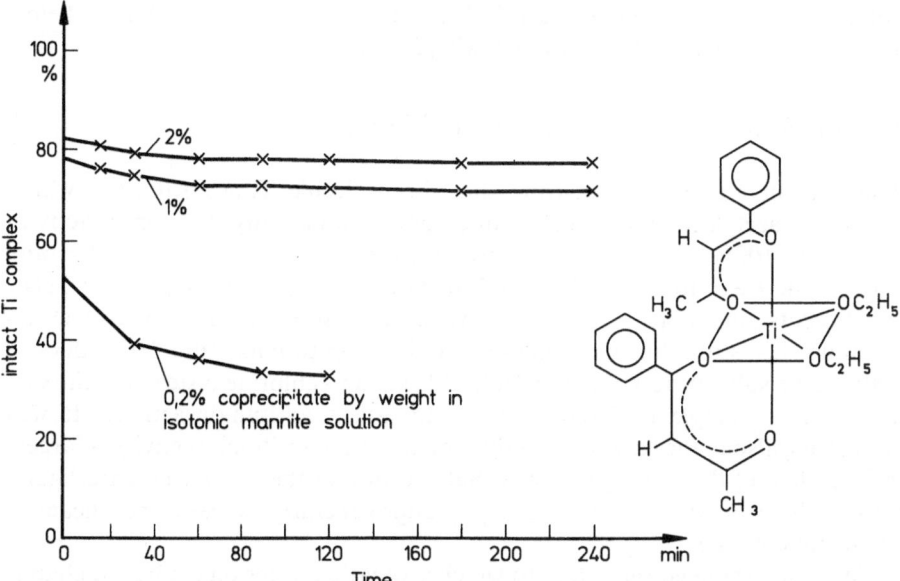

Fig. 9. Time-dependent decomposition of budotitane in an aqueous micellar solution of the coprecipitate. a) 2%; b) 1%; c) 0.2% coprecipitate by weight in isotonic mannite solution

hours in the region from 1 to 2 percent coprecipitate in weight in isotonic mannite solution (or physiological saline). The initial rapid decomposition must probably be attributed to the destruction of micelles that were defective from the start. When smaller amounts of coprecipitate are dissolved, less than 0.3 percent in weight, the formation of micelles is obviously disturbed, and a more pronounced decrease in intact titanium complex to about 40% is found within two hours (Fig. 9). The coprecipitate solutions used in the clinic have the favourable concentration range of 1 to 2 percent of coprecipitate in weight. This corresponds to about 100 to 200 mg of budotitane per 100 ml [17, 53].

8 Structure–Activity Relation of Tumor-Inhibiting Bis(β-diketonato) Metal Complexes

The relation between chemical structure and antitumor activity of bis(β-diketonato) metal complexes will be described in the following on the basis of some characteristic representatives chosen from about 200 complexes synthesized and examined. These representatives are presented in Tables in the order of increasing activity in the sarcoma 180 ascitic tumor model. The T/C values in percent (median survival time of treated animals versus median survival time of control animals ×100) indicate antitumor activity. In an active compound, T/C values should be high, that is, they should exceed 125%. If a compound attains a value between 130% and 100%, it is inactive. Values below 100% mean that the compound is toxic [25, 27, 55].

8.1 Variation of the β-Diketonato Ligand

The data in Table 2 demonstrate that the molecule $Ti(acac)_2(OEt)_2$, which bears the acetylacetonato ligand, does not produce any antitumor activity (T/C = 90–100%). When a methyl group is replaced by a tertiary butyl group, activity increases to T/C values of 130–170%. The inclusion of other bulky substituents–in Table 2 there is a cyclohexane-substituted derivative as an example–leads to a further increase in activity up to maximum T/C values of 200%. Sole substitution of the hydrolizable group–chloride instead of ethoxide as the last example in Table 2–does not affect antitumor activity. It was confirmed, also in the case of other diketonates, that antitumor activity is largely independent of the leaving group X. Substitution of the acetylacetonato ligand for lipophilic, bulky alkyl or cycloalkyl groups generally increases the efficacy of the metal complex.

We also examined the extent to which aromatic groups on the ligand change the efficacy of the complexes. The data are summarized in Table 3. We compared the effect of phenyl substituents on antitumor activity of the compounds

Table 2. Antitumor activity of titanium complexes of the type Ti(dik)$_2$X$_2$ in the sarcoma 180 ascitic system at a single dose of 0.2 mmol/kg of metal complex given 24 hours after tumor transplantation. As diketonate ligand we used acetylacetone and compounds derived from it with space-filling aliphatic substituents. T/C(%) = (median survival time of treated animals vs median survival time of control animals) × 100

β-diketonate	X	T/C (%)
(acetylacetonate structure)	OEt	90 – 100
(tert-butyl substituted diketonate structure)	OEt	130 – 170
(tert-butyl, cyclohexylmethyl substituted diketonate structure)	OEt	150 – 200
(tert-butyl, cyclohexylmethyl substituted diketonate structure)	Cl	150 – 200

with that of Ti(acac)$_2$(OEt)$_2$, which reaches T/C values between 90 and 100%. When a hydrogen atom of a methyl group is replaced by a phenyl group, antitumor activity increases at about the same level as when introducing a bulky alkyl group (see Table 2). If, however, the phenyl group stands in direct conjugation to the metal enolate ring, the result is a surprisingly high increase in antitumor activity. For example, the compound with a phenyl group in 2-position attains T/C values between 200 and 250%, and the derivative with a phenyl group in 1-position reaches T/C values > 300%.

One of the two highly active compounds is budotitane (no. 4 in Table 3), which is currently undergoing clinical trials. The phenyl group of the benzoylacetonato ligand hence affects the overall structure of the metal complex in such a way that an inactive complex turns into a highly active substance.

Similarly interesting effects can be obtained with the chloride derivative (No. 5 in Table 3). This is evidence that the hydrolizable group has little influence on antitumor activity. However, the ethoxy group makes for much better stability to hydrolysis. Thus the corresponding diethoxy complex has to be preferred to the chloride derivative in practice, as far as galenic formulation is concerned.

It might be assumed now that systematic variations at the phenyl ring of budotitane would optimize its activity even more. We examined the influence of

Table 3. Antitumor activity of Ti(dik)$_2$X$_2$ complexes in the sarcoma 180 ascitic tumor at a dose of 0.2 mmol/kg given 24 hours after tumor transplantation. As diketonate ligand we used acetylacetone and corresponding phenyl-substituted derivatives. T/C(%) = (median survival time of treated animals vs. median survival time of control animals) × 100

β-diketonate	X	T/C (%)
(acetylacetonate)	OEt	90 - 100
(benzyl-substituted diketonate)	OEt	130 - 170
(phenyl-substituted diketonate)	OEt	200 - 250
(benzoylacetonate)	OEt	> 300
(benzoylacetonate)	Cl	> 300

substituents with negative and positive inductive and mesomer effects on antitumor activity. Table 4 shows some characteristic derivatives and their T/C values. Obviously, the introduction of methyl groups at the phenyl ring does not alter antitumor activity, whereas methoxy, chlorine, and nitro groups reduce it. Thus antitumor activity cannot be increased this way.

All data suggests that *unsubstituted* aromatic ring systems in the periphery of the molecule have significant positive effects on the antitumor activity of such metal complexes, while substitutions at this site may be considered to be detrimental to the activity.

8.2 Variation of the Central Metal

The central metal of the complexes is another important parameter in optimizing these drugs. The activities of the titanium and the zirconium derivatives with the benzoylacetonato ligand that produces the highest level of antitumor activity are relatively similar, but they then decrease markedly in the order

Table 4. Antitumor activity of Ti(dik)$_2$Cl$_2$ complexes in the sarcoma 180 ascitic tumor at a dose of 0.2 mmol/kg given 24 hours after intraperitoneal tumor transplantation. As diketonate ligand we used benzoylacetone and its phenyl-substituted derivatives. T/C (%) = (median survival time of treated animals vs median survival time of control animals) × 100

β-diketonate	T/C (%)
	> 300
	> 300
	150 - 200
	150 - 200
	100 - 120

Hf > Mo > Sn > Ge. The germanium compound is virtually inactive even at other dosages, but the tin compound shows a minor increase in activity at lower doses. T/C values around 200% are reached at 0.1 mmol/kg. However, the therapeutic range is considerably lower than in the case of the corresponding titanium compound. The molybdenum complex, which is slightly active in the sarcoma 180, has rather a stimulating effect on tumor growth in the acetoxy-methylmethylnitrosamine-induced colorectal tumors of the rat. This is quite in contrast to the excellent effect of the corresponding titanium compound on these tumor models [16, 22].

If the homologues zirconium and hafnium are used instead of titanium, it is also possible to synthesize hepta-coordinated compounds of the type Zr(bzac)$_3$Cl and Hf(bzac)$_3$Cl. These are also active, although at considerably higher doses than the corresponding octahedrally coordinated complexes. It has to be investigated whether these complexes act in the organism directly as seven-coordinated molecules, or only after they have been converted into the corresponding octahedrally configurated, six-coordinated complexes [24].

Table 5. Antitumor activity of $Ti(dik)_2X_2$ complexes with different leaving groups in the sarcoma 180 ascitic tumor at a dose of 0.2 mmol/kg, applied 24 hours after intraperitoneal tumor transplantation. T/C (%) = (median survival time of treated animals vs median survival time of control animals) × 100

β-diketonate	X	T/C (%)
	F	90 - 100
	Cl	90 - 100
	Br	90 - 100
	OEt	90 - 100
	F	> 300
	Cl	> 300
	Br	> 300
	OEt	> 300

8.3 Variation of the Group X

The nature of the leaving group X does not seem to contribute much to the antitumor activity of this class of substances. Ethoxide, chloride, bromide, and fluoride with benzoylacetone as ligand show excellent anticancer activity. The same derivatives do not show any activity when they have acetylacetone as diketonate ligand (see Table 5). The four derivatives of the type $Ti(acac)_2X_2$ with X = F, Cl, Br, or OEt are inactive with T/C values between 90 and 100%. In contrast to this, the four $Ti(bzac)_2X_2$ derivatives reach T/C values up to 300%. They are thus highly active, irrespective of the hydrolizable group X. However, galenic behaviour is considerably influenced by this leaving group, because stability in water clearly increases in the order iodine < bromine < chlorine < fluorine < OR. Thus budotitane, rather than other, analogous compounds, was chosen for further development. The iodine compound is too unstable to be considered for further development, and the bromine and fluorine compounds, apart from galenic disadvantages, have disadvantages over ethoxide as to the way in which the hydrolizable group is physiologically tolerated.

9 Antitumor Activity in Transplantable Tumor Models

Antitumor activity in transplantable tumor models is summarised in Table 6. In this Table we must distinguish between T/C values calculated on the basis of survival time and T/C values calculated on the basis of tumor weight (cf. Sect. 8). If the evaluation parameter is tumor weight, T/C values are calculated as median tumor weight of treated animals versus median tumor weight of control animals ×100. The resulting percentage should be low for an active compound, that is, it should fall below 45%. This is quite in contrast to the T/C values based on survival time, in which case the percentage should be high (see Sect. 8).

In the Stockholm ascitic tumor, the Ehrlich ascitic tumor, and in the MAC 15A colon tumor, a transplantable colon adenocarcinoma, budotitane therapy led to T/C values of >300%. In the Walker 256 carcinosarcoma, T/C values of 200% were reached. The subcutaneously transplanted sarcoma 180 can be cured with intravenous budotitane therapy. Intravenous application also reduces the

Table 6. Survey of the most important results of budotitane therapy in transplantable tumors. T/C values >300% mean that a high percentage of animals are cured. T/C(%) = (median survival time or median tumor weight of treated animals vs median survival time or median tumor weight of control animals) × 100. ST = median survival time; TW = median tumor weight

tumor model	evaluation parameter	optimum T/C value (%)
Sarcoma 180 ascitic tumor	ST	> 300
Sarcoma 180 tumor, subcutaneously growing	TW	0
Sarcoma 180 tumor, intramuscularly growing	TW	30
Walker 256 carcinosarcoma	ST	200
P 388 leukemia	ST	130
Stockholm ascitic tumor	ST	> 300
Ehrlich ascitic tumor	ST	> 300
MAC 15A colon tumor	ST	> 300
AMMN-induced colorectal tumors	TW	20

ST = survival time; TW = tumor weight; AMMN = Acetoxymethylmethylnitrosamine; T/C(%) = (median survival time or median tumor weight of treated animals vs. median survival time or tumor weight of control animals) × 100. Activity (parameter survival time): T/C > 125%. Activity (parameter tumor weight): T/C < 45%

tumor weight of the intramuscularly growing sarcoma 180 to 30%, compared to the control experiment (100%).

It is interesting that after a primary screening in the leukemias P 388 or L 1210, budotitane would not have been considered further, because it is only marginally active in these models (T/C = roughly 130%). The quick-growing leukemias are certainly not the adequate model for finding substances that are active in slow-growing tumors such as colon tumors. However, it is the slow-growing tumors, of all cancers, which pose the biggest problem in cancer therapy today.

All tumor models that have been described in this article are transplantable tumors. This means that they can be transplanted from one animal to another within a particular strain. Positive results obtained in these experimental models indicate antitumor activity in general. If a compound is active against several of these models, it may also be supposed to be active in the clinic. The most important disadvantage of these models is that they cannot ultimately define the organ tumors against which a new compound will be active in humans, because they are not comparable to human tumors of specific origin. Therefore it is necessary to evaluate the activity of a new compound in autochthonous tumors, which are mainly induced by a carcinogen, and which mimic the human situation fairly closely owing to the characteristics given below:

1. Tumor origin, growth and therapy take place in the same individual.
2. There is genuine tumor histology.
3. The tumor–host interaction is identical with that in humans.
4. The proliferating fraction of tumor cells is lower than in transplanted tumors.
5. Tumor growth is orthotopical, i.e. autochthonous tumors do not grow at artificial sites (e.g. subcutaneously).
6. Autochthonous tumors show a lower sensitivity to conventional cytostatic agents than transplanted tumors.

These experimental models are highly developed and costly and cannot be used as a primary screen but only for compounds which have already shown antitumor activity in a number of transplantable tumors or others. Thus, after screening budotitane in transplantable tumors, we went on to investigate its antitumor activity in autochthonous tumors.

10 Therapy of Autochthonous, AMMN-induced, Colorectal Tumors with Budotitane

In our experiments we selected AMMN (acetoxymethylmethylnitrosamine)-induced colorectal tumors, which are highly predictive for the clinical situation

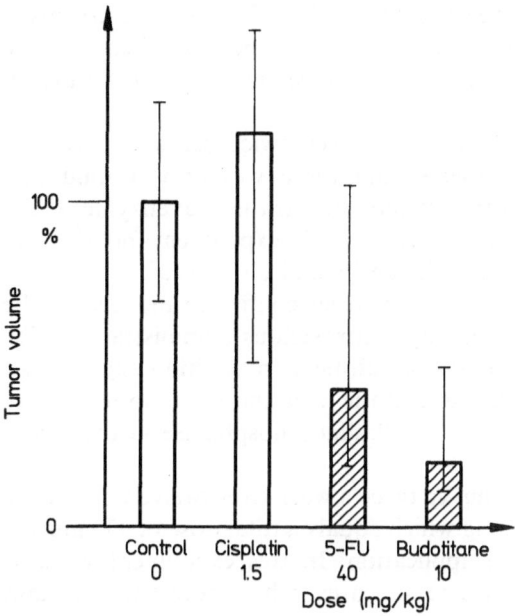

Fig. 10. Comparison of the antitumor activity of cisplatin, 5-fluorouracil, and budotitane in autochthonous, AMMN-induced colorectal tumors of the rat. The compounds listed were applied twice per week over 10 weeks after tumor manifestation. The reduction of tumor volume, which is represented by the *shaded columns*, is statistically significant, in comparison with the control group (Kruskal Wallis Test)

and which represent a tumor type that is among the most frequent causes of death from cancer, as it does not sufficiently respond to clinical chemotherapy.

Figure 10 compares the activity of 5-fluorouracil, cisplatin, and budotitane in AMMN-induced colorectal tumors. Budotitane is markedly more active than 5-fluorouracil. It reduces tumor volume to about 20% of the initial value. 5-Fluorouracil effects a tumor remission of up to 40% of tumor volume, whereas cisplatin, with a value of about 120%, stimulates tumor growth a little. Stimulating effects are not infrequent with inactive compounds.

The experiments in autochthonous colon tumors show that budotitane may have an interesting clinical indication. This is especially promising because the colon tumors are not very sensitive to established clinical chemotherapeutic agents.

11 Toxicity of Budotitane

When budotitane is given to female Sprague-Dawley rats in a single intravenous administration, the LD_{50} – that is the dose at which 50% of animals die – is at about 80 mg/kg. In mice the LD_{50} is about twice as high. Given i.v. application, rats and mice will tolerate somewhat more than twice the maximum tolerated i.v. dose.

In the course of these experiments we also detected the most important side-effects. The dose-limiting factor for budotitane is hepatotoxicity, with multiple focal necroses of the liver at about the level of the LD_{50}. There were also signs of a certain lung toxicity owing to hemorrhagic pleural effusions and hemorrhagic oedematous districts in the lung, which, however, were only found at the level of the highest lethal doses.

Chronic doses between 10 and 20 mg/kg, given twice per week over ten weeks, were tolerated without problems, and there was only a mild and reversible liver toxicity. Laboratory parameters such as the liver enzymes GOT and GPT, as well as LDH, were increased. In this experiment, no signs of myelosuppression could be observed in the peripheral blood [25].

Independent investigations in other laboratories confirmed the level of the LD_{50} to be at about 60–70 mg/kg in single intravenous administration. The other toxicological parameters were also confirmed. In addition, signs of an extramedullar blood formation were found in the liver and in the spleen. There was mild nephrotoxicity, and, in addition, alkaline phosphatase in the serum was increased.

In chronic application, up to 18 mg/kg twice a week over twelve weeks were tolerated without mortality, something which equals a total dose of 432 mg/kg, i.e. seven times the LD_{50} in single application. In this experiment we also observed a marked increase in creatinine and urea, which points to a certain degree of nephrotoxicity.

Experiments with pigeons, which have proved to be a predictive model for emesis in humans, did not reveal any emetic potential for budotitane [25]. Mutagenicity of budotitane was screened by means of the *Salmonella typhimurium* mammalian microsome assay of Ames. There were no signs of a mutagenic potential [56].

Toxicological studies with budotitane, particularly in chronic application, show that the substance is well-tolerated at levels which can be considered for therapy. Mild, reversible lever toxicity is a prominent feature. Nephrotoxicity seems to play a role only at considerably higher doses. The lack of myelosuppression is another advantage of budotitane [17, 27].

12 Budotitane Clinical Phase I Study

In the first part of a clinical phase I study with budotitane, only single applications were investigated, and in the second part budotitane was administered twice a week over four weeks. Seven different doses were applied in single application, namely 1, 2, 4, 6, 9, 14, and 21 mg/kg of bodyweight [18, 55, 57].

First signs of drug toxicity were seen at 9 mg/kg, when a patient complained about an impairment of the sense of taste shortly after the infusion. This side-effect continued to be present even at higher doses, and finally a complete loss of taste was observed. This impairment, however, was entirely reversible in all cases and was present only for a few hours after budotitane infusion.

From a dose of 14 mg/kg onwards, a minor increase in liver enzymes and in lactate-dehydrogenase was observed. From a dose of 21 mg/kg onwards, a dose-limiting nephrotoxicity was found with a rise in urea and creatinine, which was in accordance with grade 2 toxicity on the basis of WHO criteria. This side-effect was also entirely reversible after some weeks, and urea and creatinine levels returned to normal. Signs of myelotoxicity could not be observed at any dose.

The maximum tolerated budotitane dose thus ranges between 14 and 21 mg/kg. This equals about 30% of the LD_{50} in animal experiments, and this is surprisingly high.

In pharmacokinetic investigations blood was taken from the patients 10 minutes, 1, 2, 8, and 24 hours as well as seven days after budotitane infusion. The titanium level in the serum and in the erythrocytes was determined by means of atomic absorption spectroscopy. Given 14 mg/kg budotitane, the highest titanium levels in the serum were found two hours after the infusion – between 2 and 5 μg/g, and levels between 5 and 12 μg/g were found at a dose of 21 mg/kg. Titanium could be detected in the serum even after seven days and on one occasion even after four weeks. In the erythrocytes, titanium was found at a concentration between 1 and 2 μg/g.

After single application studies, side-effects of chronic budotitane application were screened. Proceeding from a maximum tolerated single dose that is slightly below 21 mg/kg, a total dose of 21 mg/kg (about 800 mg/m²) was fixed for repeated applications, which was divided into 8 applications of 100 mg/m², 120 mg/m², and 150 mg/m², twice a week over four weeks. The unit m² is used to measure the body surface area and is calculated on the basis of bodyweight and height, e.g. a person 1.80 m tall and weighing 80 kg has a body surface area of about 2.02 m². This unit makes for a better correlation between toxicity and applied dose under clinical conditions than the unit mg/kg. On the basis of the scheme described above, three patients have so far been treated at each dose level without any dose-limiting toxicity. However, all patients had the same changes in the sense of taste [17, 18, 27].

The overall picture of side-effects is one of great resemblance to that of preclinical toxicology studies. The clinical phase I study also confirmed the prediction that budotitane, unlike cisplatin, would not cause emesis. On the basis of our experiments with pigeons this could be expected. The lack of this side-effect is of the utmost importance for the compliance of the patient. Intensive vomiting, as so often occurs with cisplatin therapy, has frequently led to patients disrupting therapy on their own account. As had been predicted from preclinical studies, the main side-effects turned out to be liver and nephro-toxicity. Mild and reversible hepatotoxicity appears even at low doses, while nephrotoxicity cannot be observed unless the highest doses are applied. In case this side-effect should play a role in repeated applications – there has not as yet been any indication, one might reduce it by means of a prophylactic hyper-hydration, along with diuresis, comparable to the measures taken in cisplatin therapy.

On the whole, the spectrum of side-effects of budotitane conveys a positive impression. A subsequent phase II study will reveal the extent to which preclinical expectations of the activity of budotitane in adenocarcinomas of the gastrointestinal tract and other tumors can be confirmed.

13 Conclusion

The development of cisplatin constitutes the first real important contribution to the chemotherapy of cancer to come from the field of inorganic chemistry. Today this compound is among the most-sold anticancer drugs. Although cisplatin is widely used, it is effective only against very few tumors. Thus there may be cures in cases of testicular carcinomas, which are seen mainly in young males. In the Federal Republic of Germany, there are about 1500 cases annually. These patients can usually be cured, thanks to the introduction of cisplatin into combination chemotherapy. Cisplatin also shows relatively good activity in cases of ovarian carcinomas, tumors of the head and neck, bladder tumors and bone tumors. However, in these cases the drug is far from being as effective as it is in cases of testicular carcinomas, where one might already speak of a very selective therapy.

Numerous new platinum derivatives have been developed in order to successfully treat other tumors, but all new cisplatin analogues are very similar to the parent compound in their spectrum of indication. Thus there was no decisive breakthrough in cancer chemotherapy. A logical consequence of this was to try and synthesize compounds with other central metals. Among transition metal complexes, budotitane, *cis*-diethoxybis(1-phenylbutane-1,3-dionato)titanium(IV), is the first compound to have qualified for clinical studies after the platinum complexes [55, 58]. As could be expected, the structure–activity relation of this titanium compound is completely different from that of platinum complexes. Although we also deal with *cis*-configurated complexes here, activity depends on the β-diketonato ligand. It appears that a planar aromatic group in the β-diketonato ligand is responsible for activity. If we take a look at the same complex with acetylacetone as ligand on the one hand and with benzoylacetone as ligand on the other, we see that the acetylacetonato complex is virtually inactive while the benzoylacetonato complex is highly active. Such situations are frequently seen with anticancer drugs when activity is determined by an intercalating mechanism. Although a certain increase in antitumor activity can also be observed when bulky alkyl or cycloalkyl groups are introduced into the molecule, the effect is far from being as good as in the case of planar aromatic systems. It is interesting to note that the organometallic titanium cyclopentadienyl compounds, which were examined by another research group, show similar aspects, that is, planar systems in cases capable of intercalation [19].

The *cis*-configurated hydrolizable groups in the budotitane molecule have relatively little influence on its antitumor activity, but they are decisive in the hydrolysis of the complex and thus also in what is known as the galenic formulation of the drug for use in the clinic. Since all β-diketonato compounds as such are fairly sensitive to hydrolysis, an ethoxy group was introduced, which is slow to react and which renders the compound stable enough for application in the clinic. A difficulty in the way of the clinical administration of the drug was, of course, water-solubility, and this could only be overcome by using galenic adjuvants and emulsifying agents. Chemical derivatisation such as the introduction of groups which increase the hydrophilicity did not lead to satisfactory results. In order to achieve water-solubility it is necessary not only to replace the leaving groups by hydrophilic ligands but also to substitute the diketonato ring with hydrophilic groups, e.g. $-SO_3H$ groups. However, this procedure results in a loss of antitumor activity. It is a problem which arises quite frequently in the development of tumor-inhibiting metal complexes. If the chemical structure of an active, insoluble complex is changed in such a way that water-solubility is obtained, antitumor activity of the complex will often be lost.

The tumor-inhibiting activity of budotitane is also dependent on the central metal. A significant loss in activity is observed from titanium, zirconium, hafnium, and germanium to tin. The stereochemistry of the complexes, which is relatively similar, should play a minor role here. It is more likely that differences in the reactivity of the complexes are more important in this connection, also in terms of the hydrolysis of the diketonato group.

On the whole, budotitane shows a sophisticated and graded structure–activity relation, such as may be expected from drugs with specific activity. Budotitane has good activity against colon tumor models, and this gives us reason to hope for a new management of these tumors, which so far have not been treatable with chemotherapy.

As far as the development of other metal complexes is concerned, promising results have been obtained in the field of tumor-inhibiting ruthenium complexes. An example is trans-$HIm(RuIm_2Cl_4)$, which has shown very good activity in experimental colon tumors [8]. Ruthenium compounds with DMSO as ligand are also under preclinical studies [9]. Some interesting aspects can be found in the chemistry of tin [11–13]. However, with tin compounds there is a certain contrast between high in vitro and low in vivo activity. This indicates a fairly non-specific cytotoxicity.

Gold compounds, playing as they do a crucial role in antiarthritic therapy, have turned out a number of interesting results in the chemotherapy of cancer, but a particular indication for the clinic has not been found yet [1, 15].

Other developments proceed from the idea of linking anticancer agents such as cisplatin with carrier molecules, which will selectively transport the drug to the tumor site and thus produce more specific activity. In this connection, compounds which are linked with hormones or hormone-like structures in order to achieve a specific affinity to hormone receptor positive tumors such as mammary carcinomas or tumors of the prostate are highly interesting [6]. Another idea is the synthesis of platinum complexes which will selectively accumulate in the bone and which may be used to treat bone tumors and bone metastases. The accumulation in the bone is achieved by coupling the platinum complexes with phosphonic acids as carrier molecules [7].

Basically, one of the most important problems in the development of new tumor-inhibiting metal complexes is of a strategic nature. It consists of the adequate use and evaluation of the available models and their improvement. In the literature we often find optimistic misinterpretations on the basis of single results. At present it seems unlikely that the valuable autochthonous tumor models can be simplified so that they might be used for a broader screening.

On the whole, the chemotherapy of cancer should not be regarded with too critical an eye, as particular leukemias, Hodgkin lymphomas, testicular carcinomas, and some other, relatively rare, tumors can be cured by chemotherapeutic agents to a high percentage today. Although this stands in contrast to the relative inefficacy of these drugs in tumors which account for the major share of cancer mortality today, it shows that it is possible in principle to develop chemical substances that are capable of selectively curing particular tumors, without any inacceptable toxicity. Cisplatin and its outstanding activity in testicular carcinomas is a case in point. Why should it not be possible in future to synthesize other drugs, also from the field of inorganic chemistry, which exhibit good antitumor activity, and especially so in the most common cancers?

Acknowledgements: The authors would like to express their gratitude to the Deutsche Krebshilfe, Dr. Mildred-Scheel-Stiftung for Krebsforschung, Bonn, for the support they received during this study.

14 References

1. Crooke ST, Mirabelli CK (1983) Am J Med 109: 113
2. Lewis AJ, Walz DT (1982) In: Ellis GP, West GB (eds) Progress in Medicinal Chem, vol 19, Elsevier Biomedical, New York, p 2
3. Peyrone M: Annalen der Chemie und Pharmacie, LI, 1 (1844)
4. Rosenberg B: Interdisciplinary Science Reviews, 3, 2, 134–147 (1978)
5. Rosenberg B, VanCamp L: Nature, 222, 385–386 (1969)
6. Knebel N, Schiller Cl.-D, Schneider MR, Schönenberger H, von Angerer E: Eur J Cancer Clin Oncol, 25, 2, 293–299 (1989)
7. Klenner T, Wingen F, Keppler BK, Krempien B, Schmähl D (1990) J Cancer Res Clin Oncol, 116: 341
8. Keppler BK, Henn M, Juhl UM, Berger MR, Niebl RE, Wagner FE: Progress in Clin Biochemistry and Medicine, 10, 41–69 (1989)
9. Alessio E, Mestroni G, Nardin G, Attia WM, Calligaris M, Sava G, Zorzet S: Inorg Chem. 27, 23, 4099–4106 (1988)
10. Clarke MJ: Metal Ions in Biological Systems, Ed by H Sigel, Vol 11, Metal Complexes as Anticancer Agents, Marcel Dekker, New York, 231–276 (1980)
11. Ruisi G, Silvestri A, Lo Giudice MT, Barbieri R, Atassi G, Huber F, Grätz K, Lamartina L: J Inorg Biochem 25, 229–245 (1985)
12. Crowe AJ, Smith PJ, Atassi G: Chem.-Biol Interactions, 32, 171–178 (1980)
13. Gielen M, Willem R, Delmotte A, Joosen E, Meriem A, Melotte M, Vanbellinghen C, Mahieu B,

Lelieveld P, de Vos D, Attasi G: Main Group Metal Chemistry, 12, 1, 55–72 (1989)
14. Berners-Price SJ, Mirabelli CK, Johnson RK, Mattern MR, McCabe FL, Faucette LF, Chiu-Mei Sung, Shau-Ming Mong, Sadler PJ, Crooke ST: Cancer Res., 46, 5486–5493 (1986)
15. Berners-Price SJ, Sadler PJ: Front Bioinorgan Chem, Xavier, AV (Ed), VCH Publ, Weinheim, FRG, 376–388 (1985 [Publ 1986])
16. Garzon FT, Berger MR, Keppler BK, Schmähl D: Cancer Chemotherapy and Pharmacology, 19, 347–349 (1987)
17. Keppler BK, Heim ME: Drugs of the Future, Vol 13, No 5–6, 637–652 (1988)
18. Heim ME, Keppler BK: Progress in Clin Biochemistry and Medicine, 10, 217–223 (1989)
19. Köpf-Maier P, Köpf H: Structure and Bonding, 70, 105–181 (1988)
20. Keller HJ, Keppler BK, Schmähl D: Arzneim.-Forsch./Drug Res, 32 (II), 8, 806–807 (1982)
21. Keller HJ, Keppler BK, Schmähl D: J Cancer Res Clin Oncol, 105, 109–110 (1983)
22. Bischoff H, Berger MR, Keppler BK, Schmähl D: J Cancer Res Clin Oncol, 113, 446–450 (1987)
23. Garzon FT, Berger MR, Keppler BK, Schmähl D: Cancer Letters, 34, 325–330 (1987)
24. Keppler BK, Michels K: Arzneim.-Forsch./Drug Res, 35 (II), 12, 1837–1839 (1985)
25. Keppler BK, Schmähl D: Arzneim.-Forsch./Drug Res, 36 (II), 12, 1822–1828 (1986)
26. Keppler BK, Diez A, Seifried V: Arzneim.-Forsch./Drug Res. 35 (II), 12, 1832–1836 (1985)
27. Keppler BK, Bischoff H, Berger MR, Heim ME, Reznik G, Schmähl D: ISPCC 1987, Padua; Ed by M Nicolini, Proc 5th Int Symp on Platinum and other Metal Coordination Complexes in Cancer Chemotherapy, Martinus Nijhoff Publishing, Boston, 684–694 (1988)
28. Thompson DW: Structure and Bonding, 9, 27–47 (1971)
29. Faller JW, Davison A: Inorg Chem, 6, 182 (1967)
30. Pinnavaia TJ, Fay RC: Inorg Chem, 7, 502 (1968)
31. Bradley DC, Holloway CE: Chem Comm, 284 (1965)
32. Wilkie CA, Lin GY, Haworth DT: J Inorg Nucl Chem., 40, 1009 (1978)
33. Serpone N, Bird PH, Somogyvari A, Bickley DG: Inorg Chem, 16, 9, 2381 (1977)
34. Jones jr, RW, Fay RC: Inorg Chem, 12, 2599 (1973)
35. Fay RC, Serpone N: J Am Chem Soc, 90, 5701 (1968)
36. Lingafelter EC, Braun RL: J Am Chem Soc, 88, 2951 (1966)
37. Collman JP: Angew Chem, 77, 154 (1966)
38. Calvin M, Wilson KW: J Am Chem Soc, 67, 2003 (1945)
39. Martell RE, Calvin M: Die Chemie der Metallverbindungen Verlag Chemie Weinheim (1958)
40. Bayer E: Angew Chem, 3, 325 (1964)
41. Bayer E: Angew Chem, 76, 76 (1964)
42. Collman JP, Moss RA, Maltz H, Heindel CC: J Am Chem Soc, 83, 531 (1961)
43. Kluiber RW: J Am Chem Soc, 82, 4839 (1960)
44. Kluiber RW: J Am Chem Soc, 83, 3030 (1961)
45. Collman JP, Marshall RL, Young (III) WL, Sears jr CT: J Org Chem., 28, 1449 (1963)
46. Collman JP, Marshall RL, Young WL, Goldby SD: Inorg Chem 1, 704 (1962)
47. Barker RH, Collman JP, Marshall RL: J Org Chem., 29, 3216 (1964)
48. Djoidjeric L, Lewis J, Nyholm RS: Chem Ind., 122 (1959)
49. Bock B, Flatau K, Junge H, Kuhr M, Musso H: Angew Chem, 7, 83, Jahrg, 239–249 (1971)
50. Hon P.-H, Belford RL, Pfluger CE: Inorg Chem., 5, 516 (1966)
51. Bradley DC, Holloway CE: J Chem Soc (A), 282 (1969)
52. Serpone N, Fay RC: Inorg Chem., 8, 11, 2379 (1969)
53. Krüger U, Kukat B, Keppler BK: – unpublished results–
54. Smith GD, Caughlan CN, Campbell JA: Inorg Chem 11, 2989 (1972)
55. Keppler BK, Berger MR, Klenner T, Heim ME: Advances in Drug Research, 19, 243–310 (1990)
56. Keppler BK, Heim ME, Flechtner H, Wingen F, Pool B: Arzneim.-Forsch./Drug Res, 39 (I), 6, 706–709 (1989)
57. Heim ME, Flechtner H, Keppler BK, Queißer W: Contrib Oncol, 37, Queißer, W., Fiebig, HH (Eds) Karger, Basel, 168–175 (1989)
58. Keppler BK: Nachr Chem Tech Lab, 35, 10, 1029–1036 (1987)

The Structure and Reactivity of Arsenic Compounds:
Biological Activity and Drug Design

Orla M. Ni Dhubhghaill and Peter J. Sadler

Department of Chemistry, Birkbeck College, University of London, Gordon House, 29 Gordon Square, London WC1H 0PP, United Kingdom

Arsenic is a potentially essential element for mammals and is abundant in certain marine organisms. Arsenic compounds have been widely used as drugs in the past and were used until recently as growth promoters for some farm animals. Here we review the structures, reactivities and biological and pharmacological activities of arsenic compounds, emphasising both the similarities to and differences from other Group V elements (P, Sb, Bi). There appear to be no crystal structures of As-containing peptides or proteins, and the inhibition of thiol-enzymes by As(III), although widely quoted, is poorly understood. There is much scope for the design of arsenic compounds as drugs, but progress depends on the discovery of new probes for investigating the structural and redox chemistry and biochemistry of arsenic, especially in aqueous media.

Structure and Bonding 78
© Springer-Verlag Berlin Heidelberg 1991

1 Introduction

Arsenic compounds are most famous for their toxicity, arsenious oxide being used extensively in the Middle Ages and later to get rid of opponents and enemies [1]. This continued until the advent of As-specific tests (e.g. the Marsh Test) and, later, instrumental methods of analysis (e.g. neutron activation analysis, Sect. 2.5). However, arsenic compounds have also been used medicinally for nearly 2,500 years [2–4].

With any element the toxicity will usually depend critically on the type of chemical compounds administered, its dose, the route of administration and often the biochemical status of the host. Low doses of certain compounds can be therapeutic, provide the necessary daily intake of an essential element, or be tolerated without harm if the element is non-essential, whereas high doses, even of compounds of essential elements, will be toxic. Arsenic may be an essential element for mammals, although this has yet to be convincingly proven (Sect. 3.1.1).

Hippocrates (460–377 B.C.) is reported [4, 5] to have recommended the use of a paste of realgar, As_4S_4, for the treatment of ulcers. In this century the most extensive use of arsenic compounds has been in the treatment of syphilis and trypanosomal diseases [3, 6, 7], although a variety of other conditions have been treated with As compounds including asthma, diabetes, and skin diseases [4] and until the 1950s [8], arsenic containing preparations were used as so-called tonics, e.g. Fowler's solution [9], reportedly first used by Fowler in 1786 [10]. Prolonged use of this can produce chronic arsenic poisoning [8]. Other As-containing preparations described as common in 1960 [4] include Donovan's solution (a mixture of arsenic and mercury iodides) and asiatic pills which contained As_2O_3 and black pepper! In India ayurvedic medicines such as bhasams which contain As [11] were in use as recently as ten years ago. Arsenic also is still used in Chinese herbal medicines [8].

In view of the known potent physiological properties of many arsenic compounds, it is remarkable that little use is made of them in conventional medicine today. As we shall see this appears to arise from a lack of knowledge of the molecular biochemical pharmacology of arsenic and also a lack of adequate probes for investigating the structure and bonding in arsenic compounds in the solid-state and in solution. For example, we shall find that the inhibition of enzymes by arsenic compounds is often attributed to attack on cysteine thiol groups, but we have not been able to find a single example of a crystal structure of a protein or polypeptide where such an adduct has been characterised.

Table 1. Nomenclature for selected arsenic compounds. Based on [36]

Oxidation state	Name	Compound
As(I)	Arseno-	$(RAs)_n$
As(III)	Arsenious acid	$As(OH)_3$
	Arsonous acid	$RAs(OH)_2$
	Arsinous acid	$R_2As(OH)$
	Arsine	R_3As
	Arsenoso-	$(RAsO)_n$
As(V)	Arsenic acid	H_3AsO_4
	Arsonic acid	$RAsO(OH)_2$
	Arsinic acid	$R_2AsO(OH)$

In this article we attempt to relate the chemistry of arsenic to the biological activity of its compounds and to examine critically some of the statements made about arsenic in the biochemical literature. Where possible comparisons will be made with the chemical and biological properties of the other elements in Group V (Group 15 in IUPAC nomenclature) of the periodic table. Although new nomenclature for arsenic compounds is probably called for since there are many confusing statements in the literature, we have attempted here to follow some recent precedents with the choice of terms listed in Table 1.

2 Chemistry

2.1 Oxidation States and Redox Chemistry

There is some confusion in the literature over the assignment of oxidation states of arsenic in various compounds as previously noted [12], e.g. some workers describe arsenic in $(CH_3)_3As$ as $As(-III)$ [13], whereas it is more usual to assign this as $As(III)$ [14] regarding the methyl group formally as CH_3^-. We will adopt the latter convention in this article and assign the oxidation state as $As(III)$ in R_3As, $R_2As(OH)$, $RAs(OH)_2$, $(RAs)_2O$, and $(RAsO)_n$, as well as in $As(OH)_3$. The four-coordinate species R_3AsO, H_3AsO_4, $R_2As(O)OH$, $RAsO(OH)_2$, and the five-coordinate spiroarsoranes (see Sect. 2.2) contain $As(V)$. $As(OH)_3$ can be reduced to the arsine at a Hg electrode [15] and arsenic in AsH_3 is best described as $As(-III)$. In the arseno compounds $(RAs)_n$, which are cyclic or polymeric in structure, As may be regarded formally as $As(+I)$.

Oxidation state diagrams for arsenic and phosphorus in acidic and basic solutions are shown in Fig. 1a. The potentials are calculated from the data given by Santhanam and Sundaresan [15, 16]. It should be noted that according to these data the $As(III)/As(O)$ and $As(V)/As(III)$ potentials in alkaline solution are negative and not positive as displayed in some standard texts [14, 17]. They

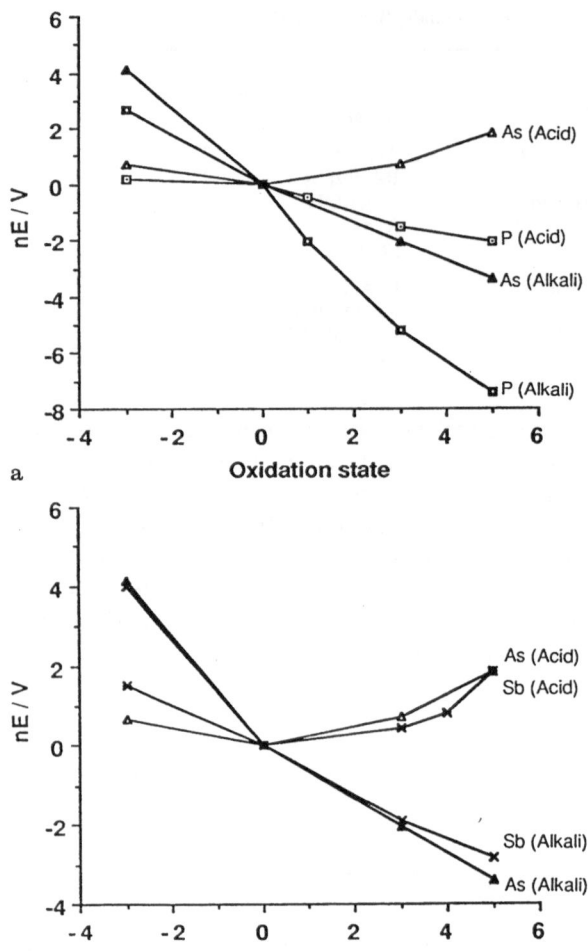

Fig. 1a, b. Oxidation state diagrams for (**a**) arsenic and phosphorus, and (**b**) arsenic and antimony. Data from [15, 16]

show that As(III) and As(V) are more easily reduced than P(III) and P(V) in alkaline solutions, Fig. 1a. The stabilities of As(III) and As(V) in acidic and alkaline solutions are comparable to those of Sb, Fig. 1b. Reduction potentials for E(V) to E(III) are substantially lower for E = P than for either As or Sb, Fig. 2 [15, 16], i.e. As(V) and Sb(V) are more easily reduced than P(V).

Evidence for the existence of other oxidation states of arsenic has been reported. Spin trapping of the products from the reaction of As(III) with HO· gives an As(IV) species [18], and As(VI) is a possible product from the reaction of $[SO_4]^-$ with As(V). A value of $E° < 1.31 \, V$ has been deduced for the $H_2As(IV)O_3/H_2As(III)O_3^-$ couple [18].

Much work has been described in the earlier literature on the pharmacologically important reduction of arsonic, $RAs(V)O(OH)_2$, and arsinic,

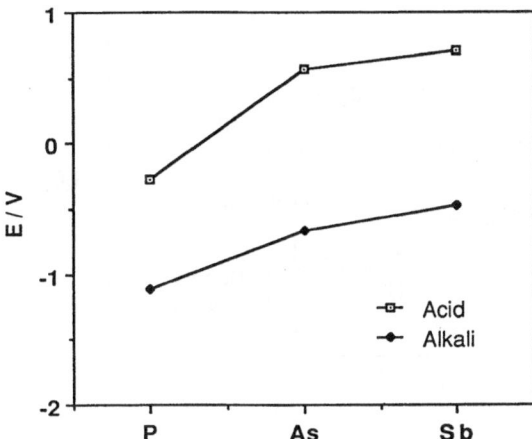

Fig. 2. Reduction potentials for E(V)/E(III) couples in acidic and basic solutions [15, 16]

$R_2As(V)O(OH)$, acids (reviewed by Doak and Freedman) [19]. The product formed depends on the reducing agent and the reaction conditions, e.g. with PCl_3 or SO_2/HI, arsonic acids yield either $RAsCl_2$ (if in acid) or $(RAsO)_n$. The action of stronger reducing agents such as H_3PO_3, $NaHSO_3$, or $SnCl_2/HI$ on $RAsO(OH)_2$ (where R is aliphatic or aromatic) gives arseno compounds $(RAs)_n$. Arsonic [20] and arsinic [19] acids have also been reduced electrolytically, e.g. the $E_{1/2}$ value for phenylarsonic acid has been polarographically determined as -0.945 V at pH < 3 [20]. Polarographic studies of arsphenamine ("Salvarsan", see Sect. 3.3.1) at the dropping mercury electrode show that the compound itself is not reduced, although unidentified impurities give rise to polarographic waves [21]. A number of other studies [20, 22] on polarographic reduction of organoarsenic compounds have been reported, e.g. arsenosobenzene $(C_6H_5AsO)_n$ [22]. The polarogram of this species was found to show up to three peaks depending on the pH of the solution and the pH dependence was also observed for the reduction when studied by cyclic voltammetry or coulometry. A mechanism has been proposed consistent with this involving the formation of the species, $(C_6H_5As)_6$, as intermediate from reduction of phenylarsonous acid, the final product being phenylarsine (gas). It has been suggested [20, 22] that arsenosobenzene forms phenylarsonous acid in solution, however this may not be the case. Furthermore, although the exact nature of the compound itself is not known, it is more likely from i.r. and molecular mass measurements to be cyclic or polymeric in the solid state and there is no evidence for the presence of an As$=$O bond, Sect. 2.5.1, so the concentrations calculated may not be accurate.

Cyclic voltammetry of arsenobenzene, $(C_6H_5As)_6$, which has a structure related to that of arsphenamine (i.e. contains As–As bonds, Sect. 2.2), has also been reported and its electrochemical behavior is similar to that of the phosphorus analogue [23]. The $E_{1/2}$ value for $(PhP)_n$ (n $=$ 5 or 6) is -2.62 V vs. Ag/Ag$^+$ [23] and the cyclic voltammetric results are consistent with the following

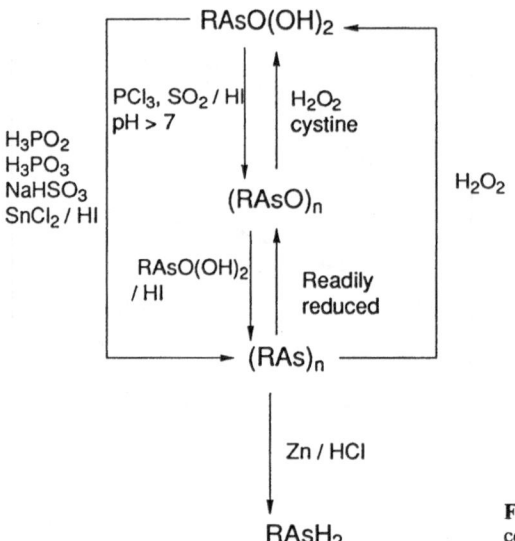

Fig. 3. Redox reactions of arseno/arsenoso compounds. Based on Ref. [19]

reactions:

$$(C_6H_5P)_n \xrightarrow{2e^-} [(C_6H_5P)_n^{2-}] \xrightarrow{fast} \text{Ring degradation products}$$

$$n = 5 \text{ or } 6$$

Some redox reactions of arseno and arsenoso compounds are shown in Fig. 3.

2.2 Solid-state Structures of Arsenic Compounds

Structurally As compounds are similar to their P analogues and where known to those of Sb. Relatively few Bi compounds are known and, in general, they are more complex structurally. The structures of Sb compounds are more varied than those of As or P. Table 2 lists the range of stereochemistries known for As and other Group V elements together with the coordination numbers [24–29].

2.2.1 As (III) Compounds

For As with an oxidation state of $+3$, as for the other Group V elements, the geometries are tetrahedral as in $[AsMe_4]I$ [30] or trigonal pyramidal, e.g. $AsPh_3$ [30]. Distortion due to stereochemically active lone pairs may be present, e.g. the dithiocarbamate complexes $[E(S_2CNEt_2)_3]$ (E = As [31], Sb [32], Bi [32]) containing six-coordinate E(III) have three long and three short E–S bonds. In the As complex these distances are 2.845 and 2.349 Å.

The As(III) oxide, As_2O_3, is commonly found as the mineral arsenolite [14]. Industrially the major source of As_2O_3 is from the flue dust of the smelting of Cu

Table 2. Structural data for Group V compounds (Based partly on [30])

No. of electron pairs	Type of hybrid	Coordination No.	No. of lone pairs	Bond Arrangement	Examples As	Others
4	sp^3	4	0	td.	$[AsMe_4]I$	$[PH_4]I$
		3	1	trig. py.	$AsCl_3$	PPh_3, $SbCl_3$, $BiCl_3$
—	—	3	—	trig. planar	$PhAs[Cr(CO)_5]_2$ [24]	$PhP[Mn(CO)_2Cp]_2$ [25] $PhSb[Mn(CO)_2Cp]_2$ [26]
5	sp^3d	5	0	trig. bipy.	$AsPh_5$	PPh_5, PCl_5, $SbPh_3(OMe)_2$, $BiPh_3Cl_2$ $SbPh_5$
				tetrag. py.[1]		
6	sp^3d^2	4	1	sq. py	$[As_2(C_4H_2O_6)_2]^{2-}$ [28]	$K_2[Sb_2(C_4H_2O_6)_2].3H_2O$ [29]
		6	0	oct.	$K[AsF_6]$	$[NH_4][PCl_6]$, $K[SbF_6]$
7	sp^3d^2	5	1	sq. py.		$[Sb_3F_{10}]^-$
		6	1	dist. oct. D_{2d}		$[SbBr_6]^{2-}$

[1] May only be a crystal packing effect, since Sb(p-tolyl)₅ is trig. bipy. [27]

[2] Compound is $[(NH_4)_4][(Sb^{III}Br_6)(Sb^V Br_6)]$

and Pb concentrates. A number of different polymorphs of arsenic trioxide are known, Fig. 4 [14]. As_4O_6 (vapour) is structurally similar to P_4O_6, the As atom in a tetrahedral environment having three pyramidal bonds [30], Fig. 5. There are two forms of As_2O_3, claudetite, [30, 33] both layer structures differing only in the way the As atoms are arranged, Fig. 6. The structure of crystalline P_2O_3 is not known.

In neutral or acid solutions arsenic trioxide is probably present as $As(OH)_3$, Sect. 2.3. The meta-arsenite salt which has been sometimes formulated as $As(ONa)_2(OH)$ or as Na_2HAsO_3 has been shown to be polymeric in the solid

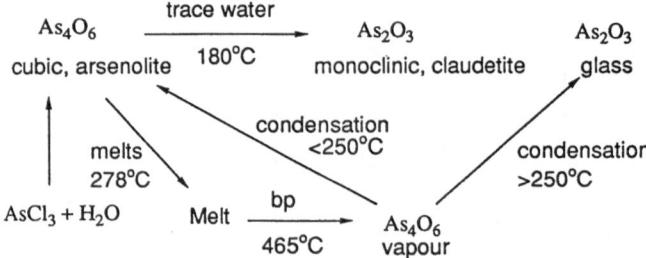

Fig. 4. Polymorphs of As_2O_3. Redrawn from Ref. [14]

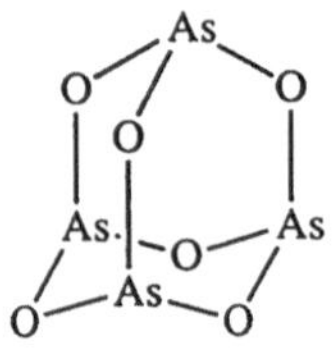

As_4O_6 **Fig. 5.** Structure of As_4O_6 in the vapour phase. Based on Ref. [30]

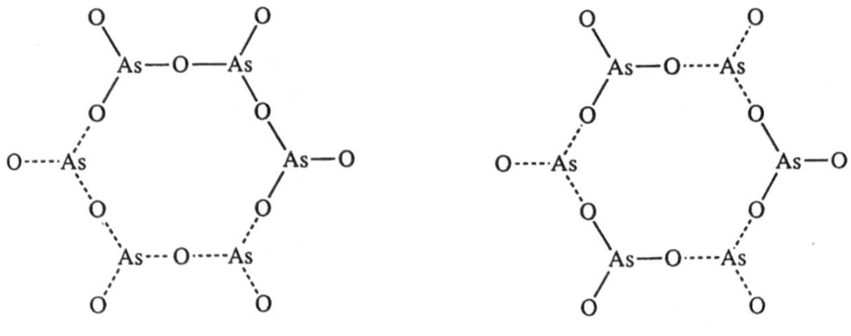

claudetite-I   claudetite-II

Fig. 6. Structures of two polymorphs of As_2O_3 (based on Ref. [33]). Arsenic atoms with *solid bonds* to oxygen (—) are above the plane of the oxygens, those with *dotted bonds* (– – –) are below

state comprising chains of pyramidal AsO_3 groups linked together by shared oxygen atoms [34], the As being at the apex of the pyramid and the three oxygens forming the triangular base. Four chains run through each unit cell held together by Na^+ ions. Thus the compound is better formulated as $(NaAsO_2)_n$, Fig. 7a. This is in contrast to the meta-phosphites such as $Mg(HPO_3).6H_2O$ in which the phosphite ion has a tetrahedral coordination, Fig. 7b, and is mono-meric with a hydrogen atom coordinated to P (P-H 1.47 Å) [35].

The structures of the Cu(II) arsenites formulated as [36] $Cu_2(C_2H_3O_2)(AsO_3)$, Paris Green, and $Cu(HAsO_3)$, Scheele's green, widely used in the late eighteenth and early nineteenth centuries as pigments in paints and wallpapers [37], do not appear to have been determined. Neither arsenious acid, $As(OH)_3$, nor the organic As(III) acids, R_2AsOH and $RAs(OH)_2$, have been crystallised. Possible solution structures are discussed in Sect. 2.3. The organo-As(III) arsenoso compounds $(RAsO)_n$ have played an important role in the development of arsenical drugs. The solution chemistry of the methyl and phenyl derivatives has been studied, section 2.5.1, and they are probably mixtures of cyclic and linear polymers [38]. However only one example of an unsupported ring of this type has been crystallised [39], $[(mesityl)AsO]_4$ (mean As–O 1.790 Å). It consists of an eight-membered As_4O_4 ring in a crown-like conformation. The methyl derivative has been crystallised only as a metal coordination complex, $[(MeAsO)_6\{(Mo(CO)_3\}_2]$ [40] (mean As–O 1.79 Å) which has a 12-membered alternating As–O ring coordinated to two $Mo(CO)_3$ groups.

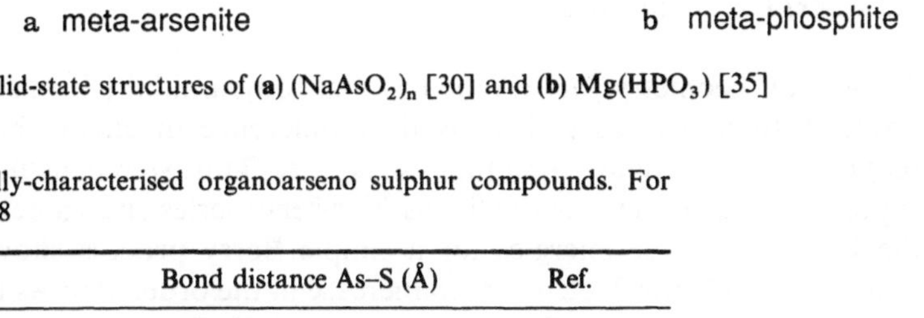

a meta-arsenite   b meta-phosphite

Fig. 7a,b. X-ray solid-state structures of (a) $(NaAsO_2)_n$ [30] and (b) $Mg(HPO_3)$ [35]

Table 3. Structurally-characterised organoarseno sulphur compounds. For structures, see Fig. 8

Compound	Bond distance As–S (Å)	Ref.
$(Me_2As)_2S_2$	2.075 (double bond)	[41]
	2.214	
	2.279	
$Ph_2As_2S_3$	2.252	[42]
	2.253	
5-Cl-1-O-4,6-S-5-arsaocane	2.249	[43]
$(PhAsS)_4$	2.262	[44]
	2.25	

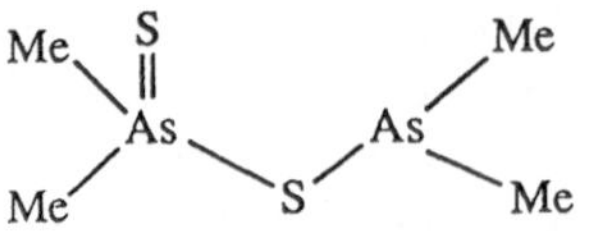

a   (Me₂As)₂S₂

b   Ph₂As₂S₃

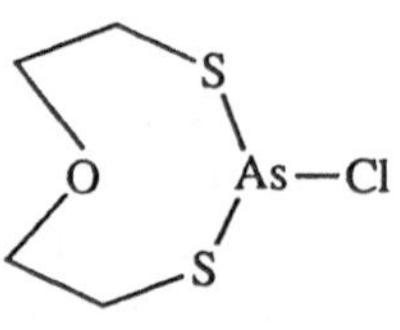

c   5-Cl-1-O-4,6-S-5-arsaocane

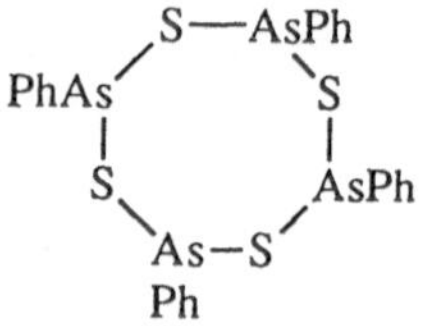

d   (PhAsS)₄

Fig. 8a–d. Structures of some arsenic compounds containing As–S bonds as determined by crystallography [41, 42, 43, 44]

Another class of compounds important in the biological chemistry of arsenic are the thioarsenites, i.e. compounds containing R_nAs(III) bonded to one or more SR groups, since the basis for the toxicity of As(III) is considered to be its interaction with thiol-containing enzymes and proteins. However only a few As–S compounds have been structurally characterised in the solid state [41–44], Table 3, Fig. 8, none of which are analogous to the proposed As–S-Enzyme complexes, Sect. 3.1.3. The cyclic structure [44] of (PhAsS)₄ is of interest in that it suggests the possibility of a similar cyclic structure for the analogous $(PhAsO)_n$.

2.2.2 As(V) Compounds

A common geometry for E(V) is trigonal bipyramidal [45] as in AsPh₅ [30], Table 2. In many cases there is little difference in energy between trigonal bipyramidal and square pyramidal structures. Thus stereochemical non-rigidity is possible, e.g. in the spirocyclic bis(biarylene) series shown below [46], Fig. 9, the NMR exchange energies for a simple Berry process show the non-rigid nature of the Group V elements to increase in the order P to As to Sb, i.e. the Sb complex has the most fluxional character. Both trigonal bipyramidal (e.g. Fig. 10a) and square pyramidal (e.g. Fig. 10b) structures have been found in crystallised As compounds [47].

Structural studies of a series of spiroarsoranes with rings of varying saturation attached to As have shown that the order of increasing displacement from trigonal bipyramidal to rectangular planar parallels the order of increasing ring delocalisation [48]. This is similar to the trends reported for phosphoranes.

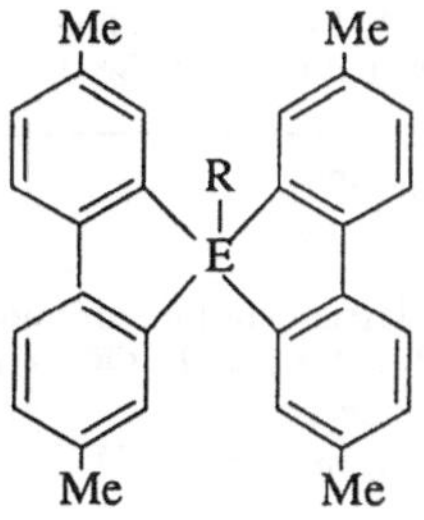

Fig. 9. Solid-state structure of fluxional Group V compounds. Redrawn from Ref. [46] R = biphenyl; E = P, As, Sb

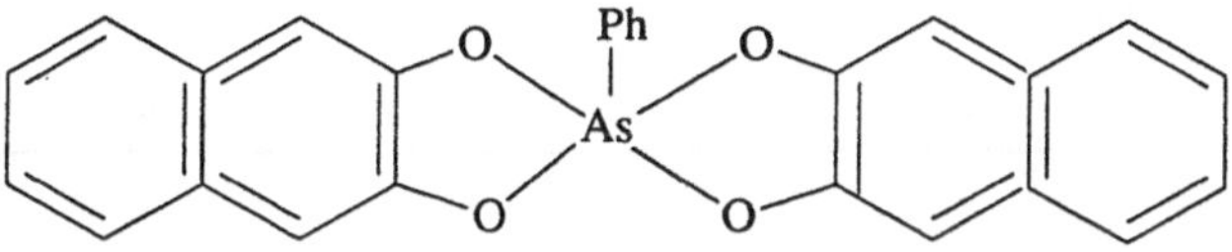

R = biphenyl ; E = P, As, Sb

a

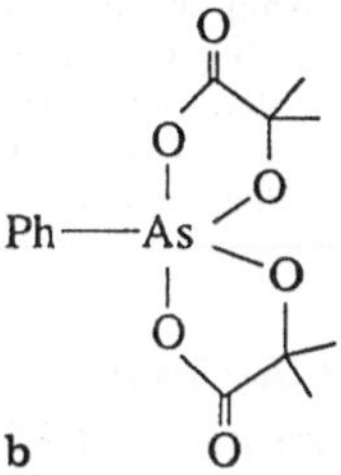

b

Fig. 10 a,b. X-ray determined structures of five-coordinate arsenic compounds (redrawn from Ref. [47]): (a) square pyramidal (b) trigonal pyramidal

The As(V) oxide, arsenic pentoxide, As_2O_5, consists of tetrahedral AsO_4 and octahedral AsO_6 groups linked in a 3D framework with As–O bond lengths of 1.68 and 1.82 Å respectively [30]. Unlike P_2O_5, As_2O_5 cannot be prepared by oxidation of the element. It is prepared by dehydration of 'hydrated' forms of the oxide which may also be regarded as hydrates of arsenic acid [30]:

$$As_2O_5.7H_2O \xrightarrow{-30\,°C} As_2O_5.4H_2O \xrightarrow{36\,°C} As_2O_5.5/3H_2O \xrightarrow{170\,°C} As_2O_5$$
$$(H_3AsO_4.2H_2O) \quad (H_3AsO_4.1/2H_2O) \quad (H_5As_3O_{10})$$

The compound, arsenic acid, $H_3AsO_4.1/2H_2O$, has been crystallised (in contrast to arsenious acid). The structure consists of a 3D arrangement of tetrahedral $AsO(OH)_3$ groups and H_2O molecules linked by hydrogen bonding [30, 49] similar to the structure of the hemihydrate of phosphoric acid, $H_3PO_4.1/2H_2O$ [50] (P–O 1.49 Å; P–OH 1.552 Å) while the compound formulated as $H_5As_3O_{10}$ is made up of infinite chains of octahedral and tetrahedral groups sharing vertices [30].

In all the arsenates, i.e. $[AsO_4]^{3-}$, $[AsO_3(OH)]^{2-}$, $[AsO_2(OH)_2]^-$, there is tetrahedral coordination of As(V) by oxygen [30] as for P(V) in the phosphates. Again unlike the organic As(III) acids, the As(V) arsonic and arsinic acids have been structurally characterised in the solid state. For both $RAsO(OH)_2$ and

Table 4. Biologically important organo-arsenic compounds characterised by X-ray crystallography. For structures. see Fig. 11

Compound	Bond distances (Å)		Ref.
	As–C	As–O	
Acetylarsenocholine bromide [(CH$_3$)$_3$AsCH$_2$CH$_2$OCOCH$_3$]Br	1.904–1.910 (As–CH$_3$) 1.933 (As–CH$_2$)		[52]
Arsenobetaine [(CH$_3$)$_3$As$^+$CH$_2$COO$^-$]	1.86–1.92 (As–CH$_3$) 1.89 (As–CH$_2$)		[53]
3-[5-Deoxy-5-(dimethylarsinoyl)-β-D-ribofuranosyloxy]-2-hydroxypropyl hydrogen sulphate	1.88, 1.91 (As–CH$_3$) 1.910 (As–CH$_2$)	1.733	[54]
p-Aminophenyl arsonic acid [(4-NH$_2$-C$_6$H$_4$As(O)(OH)$_2$]	1.95	1.73	[55]
o-Aminophenyl arsonic acid [(2-NH$_2$-C$_6$H$_4$As(O)(OH)$_2$]	1.92	1.73 (As–O) 1.64 (As=O)	[57]
3-Nitro-4-hydroxyphenyl arsonic acid [3-NO$_2$-4-OH-C$_6$H$_3$As(O)(OH)$_2$]	1.91	1.70	[56]
10-Phenoxarsine chloride	1.94	2.255 (As-Cl)	[58]

R$_2$AsO(OH), there is again tetrahedral coordination about As(V), e.g. phenyl-arsonic acid [51] and the herbicide, dimethylarsinic (cacodylic) acid [30, 51]. The latter consists of hydrogen-bonded dimers.

Few other biologically important arsenic compounds have been structurally characterised, Fig. 11 and Table 4. They include acetylarsenocholine bromide [52] and the naturally occurring arsenobetaine [53], both of which contain tetrahedral [R$_3$R′As]$^+$ groups, Fig. 11. The arsenic-containing sugar, 3-[5-deoxy-5-(dimethylarsinoyl)-β-D-ribofuranosyloxy]-2-hydroxypropyl hydrogen sulphate, isolated from the Giant Clam, section 3.4, also contains tetrahedral As(V) with an oxide and three R group ligands [54], Fig. 11. The structure of p-aminophenyl arsonic acid [55], parent acid of "Atoxyl", a former anti-trypanosomal drug (Sect. 3.3.1), again shows tetrahedral coordination about As(V) as does 3-nitro-4-hydroxyphenyl arsonic acid [56] (roxarsone, a veterinary growth promoter, Sect. 3.3.1). Only averaged As–O bond distances (As–O 1.73 Å) were reported for p-aminophenyl arsonic acid and it was not clear if there was an As=O double bond present. Subsequently the structure of the *ortho* acid [57] showed that there was indeed an As=O bond in this type of structure. The As(III) fungicide, 10-phenoxarsine chloride [58], is included as it is possibly the only biologically active organo As(III) compound which has been structurally characterised.

An unusual cyclic compound containing four and six coordinate As has been characterised [59], Fig. 12 [60].

2.2.3 As(I) Compounds

The structures of a number of As(I) compounds have been determined, e.g.   PhAs[Mn(CO)$_2$Cp]$_2$ [61],   PhAs[Cr(CO)$_5$]$_2$ [24],   and

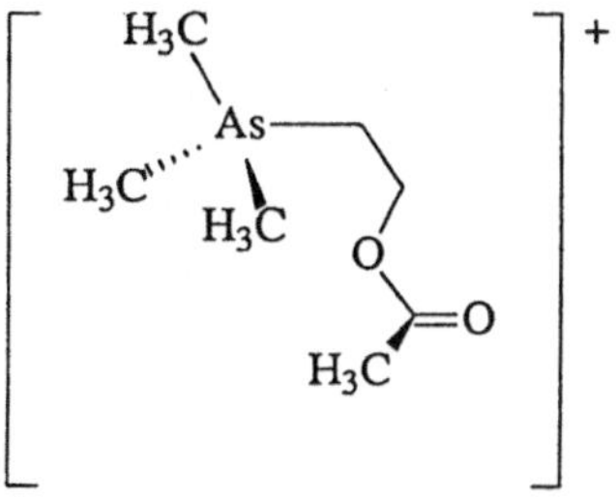

a Acetylarsenocholine bromide 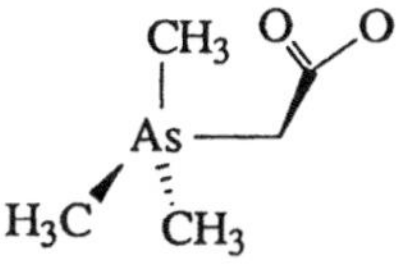   b Arsenobetaine 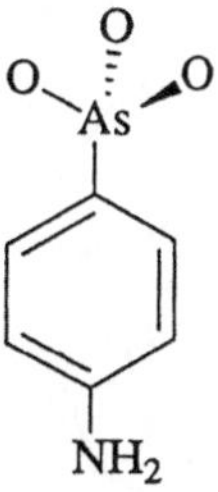   c *p*-Aminophenyl arsonic acid
                                                                        (Sodium salt is "Atoxyl")

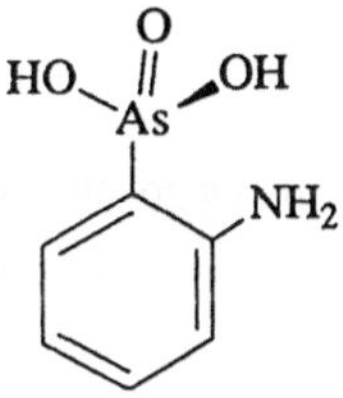

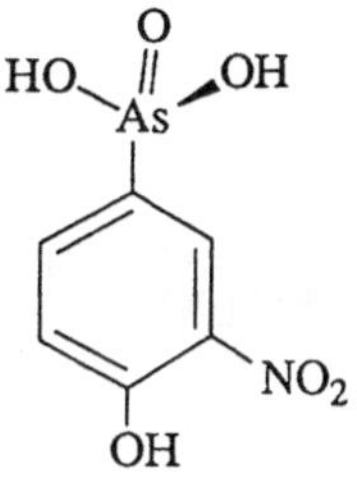

d *o*-Aminophenyl   e 3-Nitro-4-hydroxyphenylarsonic   f 10-Phenoxarsine chloride
   arsonic acid   acid (Roxarsone)   (fungicide)

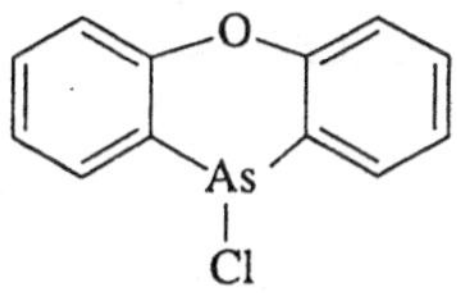

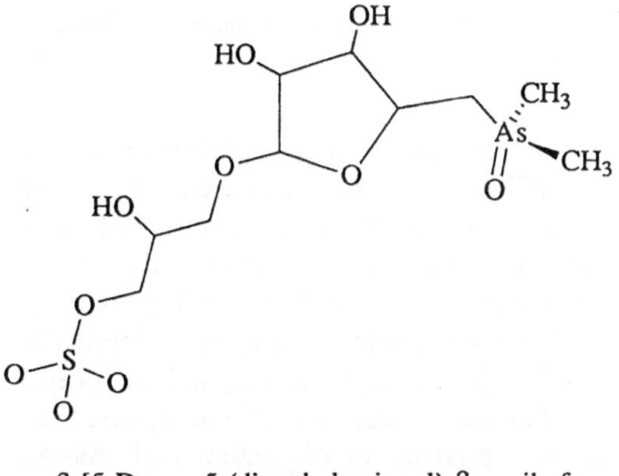

g 3-[5-Deoxy-5-(dimethylarsinoyl)-β-D-ribofuransoyloxy]-
   2-hydroxypropyl hydrogen sulphate

Fig. 11a–g. X-ray determined structures of arsenic compounds of biological interest. Redrawn from
Refs. [52, 53, 54, 55, 56, 57, 58]

$[Cr(CO)_5](Cl) As[Mn(CO)_2 Cp]$ [61]. Both P [25] and Sb [26] derivatives of
the first compound are also known. In all these compounds, the geometry about
the E atom is approximately planar as in Fig. 13.

Several catenated Group V compounds are known, e.g. cyclic compounds
such as $(RP)_n$ and $(RAs)_n$ where n can be 3 to 6, and R_2EER_2 compounds

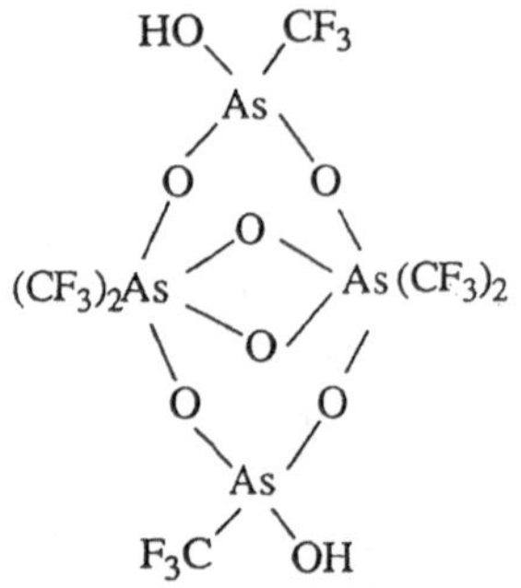

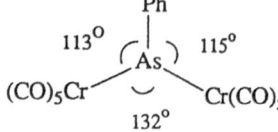

Fig. 12. Structure of a bicyclic As compound containing 4 and 6 coordinate As. Redrawn from Ref. [60]

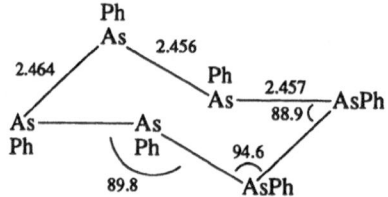

Fig. 13. Structure of a trigonal planar As(I) compound. Redrawn from Ref. [24]

Fig. 14. X-ray structure of arsenobenzene. Redrawn from Ref. [64]

(E = P, As, Sb, Bi) [62]. The solid state structures of a number of $(RAs)_n$ compounds have been determined, e.g. $(PhAs)_6$ is a six membered ring in the "chair" conformation [63, 64]. From the more recent determination [64] it is clear that of the three independent As–As bond distances, one of the bonds (2.464 (1) Å) is slightly longer than the other two (mean 2.459 Å), Fig. 14.

The original structure proposed for the drug Salvarsan invoked a kinetically unstable As=As double bond, Sect. 3.3.1 [6] as did the original proposed structure of arsenobenzene [65]. Molecular weight determinations suggest that Salvarsan is polymeric [66] containing a mixture of oligomers with As–As bonds; it appears that this compound and its derivatives are linear polymers rather than cyclic ones [65].

The factors determining the ring size of cyclic polyarsenic compounds are not clear. The number of As atoms varies depending on the identity of R, e.g. yellow $(AsMe)_n$ is a pentamer from the low temperature crystal structure [67], while the phenyl derivative is a hexamer, $(AsPh)_6$ [63, 64]. Different ring sizes with the same R group have been crystallised, e.g. both $(CF_3As)_4$ and $(CF_3As)_5$ are known [65]. Analogous cyclopolyphosphines have also been prepared [68] and again the factors determining ring size are not clear, e.g. phosphobenzene

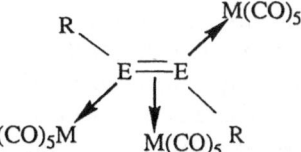

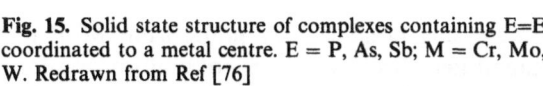

Fig. 15. Solid state structure of complexes containing E=E coordinated to a metal centre. E = P, As, Sb; M = Cr, Mo, W. Redrawn from Ref [76]

which exists in two forms in the solid state, $(PhP)_5$ and a polymorphic form, the trigonal form of which is isostructural with arsenobenzene. A correlation between endocyclic P–P–P bond angle (hence ring size) and ^{31}P chemical shift has been found [69] for polyphosphines. Electrochemical results on a number of cyclopolyphosphines [70] indicate that the potential depends on the nature of R rather than ring size, thus indicating there is little or no lone pair delocalisation in the rings. A cyclic polymeric Sb compound has recently been characterised [71], $(PhSb)_6$, isostructural with arsenobenzene. Other linear polymers of the polyarseno compounds are also known [66], e.g. for arsenomethane, the low-melting yellow $(AsMe)_5$ affords a red polymeric solid in air containing As–As single bonds [66], although this is not well characterised. There is also a crystallographically characterised purple form [72] which has a ladder-like structure consisting of 'rungs' of Me–As–As–Me (As–As bond distances 2.4 Å) [72]. Arsenobenzene also exists in a yellow linear polymeric form (not well characterised) as well as the cyclic form $(AsPh)_6$ [66]. Furthermore, in the presence of nucleophiles such as I_2 or HCl, ring-opening reactions of arseno-methane can occur affording purple-black polymeric products [65]. Since the lone-pairs on As are not delocalised around the rings, they also interact with transition metals, as reviewed by Dimaio and Rheingold [73], e.g. reaction of $(MeAs)_5$ with $[Cr(CO)_6]$ affords the cyclic compound, $[Cr_2(CO)_6\text{-}\mu\text{-}\{\eta^6\text{-}cyclo(MeAs)_9\}]$, which has a tris-homo-cubane structure.

Only recently have compounds containing As=As double bonds been isolated, the first being [74] $[(2,4,6\text{-t-}Bu_3C_6H_2)As=AsCH(SiMe_3)_2]$ with an As–As bond distance of 2.224 Å (c.f. As–As single bond distance 2.43–2.46 Å). Bulky substituents are necessary to stabilise the double bond. Similarly the only free diphosphenes known have large R groups [75]. Another method of stabilising these bonds is by coordination to a metal centre [76] and this has been achieved for P, As, Sb, e.g. Fig. 15. Indeed complexes with a formal E≡E triple bond are known [76], e.g. $[(CO)_5W]_3E_2$ (E=As, Sb, Bi) where E≡E is acting as a side-on ligand.

2.2.4 Metal Arsine Complexes

As mentioned above, metal complexes of As(I) [and P(I)] are known, however the most common are those with arsine [As(III)] ligands. Although the number of such complexes crystallographically characterised is fewer than that of the phosphines, nevertheless it is not possible to list all known complexes in a review of this type. Table 5 lists M–As bond lengths for selected metal arsine compounds together with corresponding phosphine complexes and, where known, Sb and Bi complexes [77–97].

There is a general trend of increasing bond lengths going from M–P to M–As as expected, though the magnitude of the increase is not uniform, e.g. for $[Cr(CO)_5(EPh_3)]$ going from Cr–P to Cr–As there is an increase of 0.08 Å [77], while for $[WI_2(CO)_3(Ph_2ECH_2EPh_2)_2]$ the difference from E=P to E=As is up

Table 5. E(P, As, Sb and Bi)-metal bond-lengths for selected metal Group V complexes

Compound	Metal–E bond length (Å)			Ref.
	E = P	E = As	E = Sb	
$[Cr^{(0)}(CO)_5(EPh_3)]^1$	2.422	2.4972	2.6170	[77]
$[Mo^{(II)}Cl_2(CO)_2(Ph_2ECH_2EPh_2)_2]^2$	2.593	2.681		[78]
	2.547	2.624		
	2.484	2.596		
$[W^{(III)}I_2(CO)_3(Ph_2ECH_2EPh_2)_2]$	2.508	2.680		E = P [79]
	2.495	2.605		E = As [80]
$[Co^{(0)}(NO)(CO)_2(EPh_3)_3]$	2.224	2.319	2.480	[81]
$\{[(C_4H_9)_3PCo^{(III)}(dmg_2H)$-	2.30	2.36		[82]
$As(CH_3)(C_6H_5)(O)]_2Co^{(II)}\}^3$				
$[Rh^{(I)}Cl(C_2H_4)(EPr^i_3)_2]$	2.361			[83]
	2.363			
$\{Rh^{(III)}Cl(H_2O)(EMe_3)_2[C_4(CF_3)_4]\}$		2.436		[84]
$\{Rh^{(III)}Cl(H_2O)(EPh_3)_2[C_4(CF_3)_4]\}$			2.584	[85]
			2.586	
$[Ir^{(I)}Br(CO)(tcne)(EPh_3)_2]^4$	2.397			[86]
	2.402			
$[Ir^{(I)}Cl(CO)(tcne)(EPh_3)_2]^4$		2.478		[87]
		2.481		
$[Ni^{(II)}Cl_2(Ph_2E(CH_2)_2EPh_2)_2]$	2.157			[88]
	2.145			
$[Ni^{(II)}I_2(CO)(fdma)]^5$		2.310		[89]
		2.331		
$[Pd^{(I)}(Ph_2ECH_2EPh_2)Br]_2$	2.29			
	2.26			
	2.28			
	2.32			[90]
$[Pd^{(I)}(Ph_2ECH_2EPh_2)Cl]_2(CO)$		2.45		[91]
		2.46		
$[Pt^{(0)}(C_2H_4)(EPh_3)_2]$	2.265			[92]
	2.270			
$[Pt^{(0)}(C_2F_4)(EPh_3)_2]$		2.435		[93]
		2.436		
$[Cu^{(I)}Cl(EPh_2Me)_3]$	2.289	2.36		[94]
$[Au^{(I)}Cl(EPh_3)]$	2.243			[95]
$[Au^{(I)}Br(EPh_3)]$		2.342		
$[Au^{(I)}(Ph_2E(CH_2)_2EPh_2)_2]SbF_6$	2.41			[96]
$[Au^{(I)}(Ph_2As(CH_2)_2PPh_2)_2]Cl$	2.23–2.25 (disordered, average Au-P, As)			[97]

[1] E = Bi, 2.705 Å. One of the few examples of structurally characterised Bi(III)-metal complexes
[2] Three different bond lengths observed; the equivalent P, As are compared.
[3] Co–E distances are between Co(III) and the As or P.
[4] tcne is tetracyanoethylene
[5] fdma is ferrocene-1,1'-bisdimethyl arsine

to 0.172 Å [79, 80]. As will be discussed in Sect. 3.3.3 the relative stability of M–P vs M–As may affect any biological activity of these complexes, e.g. [Au(L-L)$_2$]Cl, L-L = Ph$_2$PCH$_2$CH$_2$PPh$_2$ (dppe), Ph$_2$PCH$_2$CH$_2$AsPh$_2$ (dadpe) [97].

2.3 Acidity of Oxoacids

The pK$_a$ values for a number of As(III) and As(V) oxoacids are listed in Table 6. Data are taken from references [19, 38, 98, 99]. Also listed is the pK$_a$ of arsenobetaine (2.1) [100] which is similar to that of betaine (1.8) [101].

Esters of arsonous and arsinous acids are well-known, e.g. RAs(OR)$_2$, however the parent acids are not well-characterised [38]. Arsonous acids, RAs(OH)$_2$ may simply be hydrates of (RAsO)$_n$, i.e. (RAsO).nH$_2$O and the structures of the RAsO compounds are themselves not well known (Sect. 2.2.1). Similarly, although esters of arsinous acids, R$_2$AsOR, are known, the acids themselves may be hydrates of arsenic oxides, (R$_2$As)$_2$O.nH$_2$O. In contrast, the As(V) acids, both arsonic and arsinic, are well characterised. In general arsonic acids are stronger than carboxylic acids, but weaker than the corresponding sulfonic or phosphonic acids and weaker than arsenic acid itself (with the exception of CF$_3$AsO(OH)$_2$: pK$_1$ 1.11, pK$_2$ 5.3) [38], e.g. for PhAsO(OH)$_2$ the calculated pK$_1$ value of 3.54 shows it to be stronger than benzoic (pK$_a$ 4.2) or ArB(OH)$_2$(pK$_a$ 9.7), though weaker than ArPO(OH)$_2$ (pK$_a$ 1.84).

Table 6. pK$_a$ values for oxoacids of arsenic and arsenobetaine

Compound		pK$_a$		
Arsenious acid	"H$_3$AsO$_3$" As(OH)$_3$	pK$_1$ 9.29 [98], pK$_2$ 13.5 [109], pK$_3$ 14.0 [109]		
	H$_2$L$_2$/(HL)^2	− 0.92		
Arsonous acid	RAs(OH)$_2$	Not determined [19]		
Arsinous acid	R$_2$AsOH	Not determined [19]		
Arsenoso compounds (R = aromatic)	(RAsO)$_n$	~ 11 [19]		
Arsenic acid [98]	H$_3$AsO$_4$	pK$_1$ 2.24 ± 0.06	pK$_2$ 6.96 ± 0.02	pK$_3$ 11.50
Arsonic acid [98]	RAsO(OH)$_2$			
	R = CH$_3$	pK$_1$ 4.19	pK$_2$ 8.77	
	R = CH$_3$(CH$_2$)$_n$	pK$_1$ 4.24–4.37	pK$_2$ 9.9–9.41 (n = 2 to 5)	
Calc [19]	R = CH$_3$(CH$_2$)$_n$	pK$_1$ 4.0 to 4.5	pK$_2$ 8.9 to 9.3 (n = 2 to 5)	
	R = Ph [197]	pK$_1$ 3.47	pK$_2$ 8.48	
Arsinic acid	R$_2$AsOOH [38]			
	R = CH$_3$ (cacodylic)	pK$_1$ 1.78 [98][1]	pK$_2$ 6.14 [98]	
	R = CF$_3$	1		
	R = CH$_3$(CH$_2$)$_n$ [98]	pK$_1$ 1.48–1.69	pK$_2$ 6.43–6.54 (n = 2 to 5)	
	R = C$_6$H$_{11}$	7.25 [99]		
	R = n-C$_4$H$_9$	7.13 [99]		
Arsenobetaine		2.1 [100][2]		

[1] Refers to protonation in strong acid to form [(CH$_3$)$_2$AsO(OH)$_2$]$^+$
[2] Determined from electrophoresis

Arsinic acids can be protonated in strong acid, i.e.

$$R_2AsO(OH) + H^+ \rightleftharpoons [R_2As(OH)_2]^+$$

Lengthening of the alkyl chain in $R = CH_3(CH_2)_n$ from $n = 0–6$ has little effect on this protonation, and gives only a slight increase in basicity of $[R_2AsO_2]^-$. Cacodylic acid, $Me_2As(O)(OH)$, with a pK_2 of 6.14 is commonly used as a buffer for biochemical reactions. However, it is readily reduced by thiols (Sect. 3.1) and may well play a role in reactions of proteins containing free cysteine groups. If such proteins are crystallised from cacodylate buffers they may contain co-valently bound As atoms.

Figure 16 shows the relationship between the pKa's of As and the analogous P acids. Values for the E(V) acids are very similar, however for the E(III) acids it can be seen that arsenious is very much weaker than phosphorous acid reflecting their different solution structures. In solution there is an equilibrium between two forms, Fig. 17. It is of interest to note that the As–H bond energy is approx. 20 kcal/mol less than that of the P–H bond [102], a larger difference than is seen

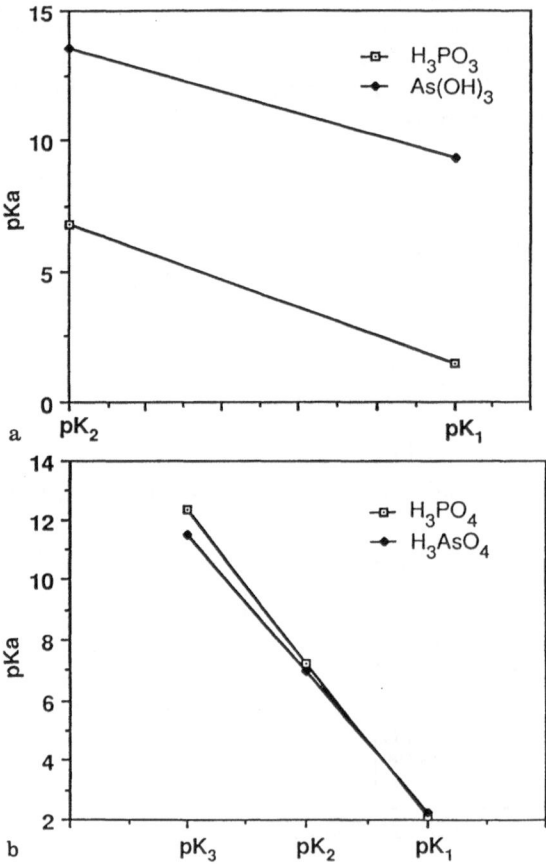

Fig. 16a,b. pK$_a$ values of (**a**) E(III) acids, and (**b**) E(V) acids (E = P, As). Data from Ref. [98]

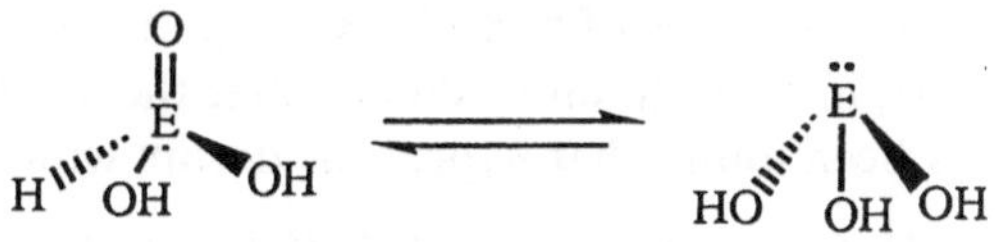

Fig. 17. Solution equilibria for E(III) acids (E = P, As)

Table 7. Trends in Group V bond lengths [30] with corresponding bond energies [102] where known

Compound	E-X (Å)			Bond energies (kcal/mol)	
	X = Cl	X = H	X = CH$_3$	X = Cl	X = H
PX$_3$	2.04	1.421	1.843	79.1	76.4
AsX$_3$	2.17	1.519	1.96	68.9	58.6
SbX$_3$	2.34	1.707	–		
	2.37				
BiX$_3$	2.50	–	2.27		

for the E–Cl bond, Table 7, and this may partly explain the position of the tautomeric equilibrium for the structure As(OH)$_3$. This equilibrium lies to the left for phosphorous acid which has a lower pK$_a$ (1.5) than phosphoric acid (pK$_a$ 2.1) and is dibasic. The H–P bond is detectable by Raman spectroscopy in both phosphorous acid [103] and in hydrogen phosphites [104] ($\nu \approx 2400$ cm^{-1}) and by ^{31}P NMR where the $\underline{P}$–H resonance appears as a doublet [105] [^1J(P–H) ca. 688 Hz (Sadler and Sutcliffe, unpublished)]. This is also the structure observed in the solid state as shown by X-ray crystallography [106]. From this there are two longer P–O bonds (P–O 1.54 Å) corresponding to P–OH and one shorter P–O bond (P–O 1.47 Å) corresponding to P=O, although the H atoms were not located. As expected from the structure only phosphites of the type, [HPO$_3$]$^{2-}$ and [HPO$_2$(OH)]$^-$ are known [30], i.e. the species [PO$_3$]$^{3-}$ is not formed. The P(OH)$_3$ species can be trapped by co-ordination to Pt, and dimerisation and dehydration of the tetrakis complex leads to the luminescent complex, {K$_2$[Pt(H$_2$P$_2$O$_5$)$_2$].H$_2$O}$_2$ [107], a P–O–P bridged Pt(II) dimer. In contrast to phosphorous acid, Raman spectra [108] of As$_4$O$_6$ in acid solution show that the major species present is pyramidal As(OH)$_3$, and no As–H vibration is observed. ^{1}H NMR spectra are reported to show [108] only a single line even at acidic pHs, thus suggesting the absence of any HAsO$_2$ species. More detailed NMR investigations of As(OH)$_3$ appear to be warranted. In basic solution Raman spectroscopy shows that the species present are [AsO(OH)$_2$]$^-$ and [AsO$_2$(OH)]$^{2-}$ with C$_s$ symmetry and [AsO$_3$]$^{3-}$ with pyramidal structures and there is no evidence of polymerisation [108]. The species, [AsO(OH)$_2$]$^-$, is similar in structure to the isoelectronic [SeO(OH)$_2$]. Arsenious acid has not been crystallised. Three arsenite anions have been implicated in kinetic studies of the reaction of arsenite (predominantly as [As(OH)$_2$O]$^-$ in alkali solution) with Ag(III) [109] to give Ag(I) and arsenate [AsO$_4$]$^{3-}$.

The kinetics of hydrolysis of a number of alkyl arsenites have been studied [110] in order to understand better the fate of arsenites in biological systems. It has been found that hydrolysis occurs with exclusive As–O bond fission [110]:

$$H_2O + As(OR)_3 \rightarrow HOAs(OR)_2 + ROH \rightarrow \ \rightarrow H_3AsO_3 + ROH$$

Results from an ^{17}O NMR study [111] on oxygen exchange between arsenite and water shows it to be a first order process probably involving nucleophilic displacement of OH by water on arsenite *via* an associative process, Fig. 18. At lower pH values, i.e. for $As(OH)_3$, the rate of exchange appeared to be faster, however due to the limited solubility of As_4O_6, this was not studied in detail [111]. It is interesting to note that the oxygen exchange of arsenate, $[H_2AsO_4]^-$, with water is catalysed by $As(OH)_3$ [112], the catalysis probably involving the formation of an arsenitoarsenate anion, $[H_3As_2O_6]^-$. This is a much slower process than for arsenite {k(arsenate exchange) 1.5×10^{-6} s^{-1} [113]; k(arsenite exchange) 167 s^{-1} [111]} reflecting the relative strengths of As(V)–O and As(III)–O bonds [111]. Knowledge of the rates of these exchanges is useful when interpreting results of studies in biological systems using O-labelled arsenite or arsenate. The hydration of CO_2 [114] is also arsenite-catalysed. Arsenite, $[As(OH)_2O]^-$, acts as a better nucleophile than arsenate, $[HAsO_4]^{2-}$, because of the greater delocalisation of charge possible in the latter [114].

Arsenate esters are more labile than the corresponding phosphate esters towards alcohol exchange [115] and hydrolysis [116]. Kinetic measurements on the rate of alcohol exchange with arsenate show this to be fast at room temperature (e.g. k for $MeOH/OAs(OMe)_3$ is 194 s^{-1} [115]) while probable reaction times of phosphate esters are hours in neutral solution at $100\,°C$. Like arsenite ester hydrolysis [110], trialkyl arsenate hydrolysis involves As–O bond cleavage [116]:

$$H_2O + OAs(OR)_3 \rightarrow OAs(OH)(OR)_2 + \ \rightarrow \ \rightarrow OAs(OH)_3 + ROH$$

Neutral hydrolysis of phosphate triesters in some cases involves C–O bond fission and is slow at $100\,°C$ so direct rate comparison is not possible, however it is clearly much slower than for arsenate hydrolysis [116]. As expected, rates of hydrolysis of arsenite esters are faster than arsenate esters, e.g. k(triisopropyl arsenate) 2.8 s^{-1} [116] and k(triisopropyl arsenite) 0.123 s^{-1} [110] under similar conditions.

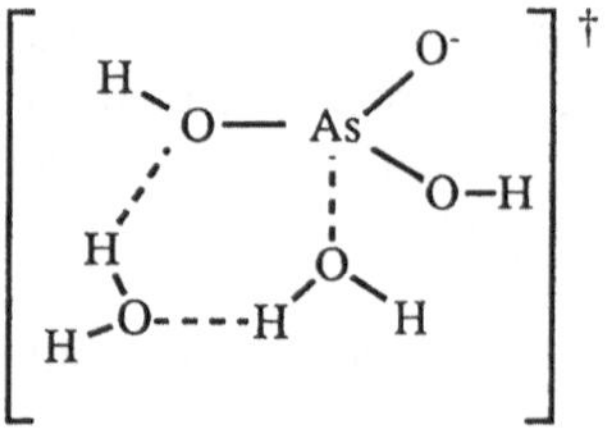

Fig. 18. Transition state involved in oxygen exchange of water and arsenite. Based on Ref. [111]

The ease of hydrolysis of arsenate esters relative to phosphate esters has important consequences for their biological activity [8]. Arsenates can replace phosphates in oxidative phosphorylation in mitochondria. However the arseno-compounds thus formed are easily hydrolysed so energy is lost rather than conserved and results in respiration occurring without phosphorylation, thus the phosphorylation is uncoupled. The half-lives of glucose-6-arsenate and 6-arsenogluconate have been calculated [117] as only approx. 6 and 30 min respectively at pH 7.0 and 25 °C, although at higher pH values and 0 °C they increase. Another case where the relative ease of hydrolysis is important is in glycolysis. Arsenate can compete with phosphate forming 1-arseno-3-phosphoglycerate which spontaneously hydrolyses forming 3-phosphoglycerate without ATP formation [118].

A series of As and Sb esters of the type $E(YR)_3$ (E = As, Sb; Y is O, S; R is alkyl or aryl) have been prepared [119], many of which are air and moisture sensitive and their NQR, vibrational and mass spectral properties investigated.

2.4 Basicity of Arsines and Related Compounds

2.4.1 Arsines and Arsine Oxides

The pK_b values for a series of Ph_3E complexes (where E is a Group V element) have been measured [120] for the reaction:

$$Ph_3E.HOAc \rightleftharpoons [Ph_3EH]^+ + [OAc]^-$$

(see Table 8). The order of decreasing pK_b values is: $Ph_3As > Ph_3N > Ph_3Bi \approx Ph_3Sb > Ph_3P$, i.e., Ph_3As is the most basic of this series of compounds. This does not correlate directly with the Group trends in electronegativity, Table 8. However a qualitative comparison of the trends in the Allred-Rochow values [121] (derived from measurements of the effective nuclear charge) and the values obtained by Sanderson [121] (from relative electron density measurements) with the above series shows that As and N are more electronegative than the other elements, hence they might be expected to have greater basicity.

Table 8. Basicity constants for Group V bases in anhydrous acetic acid (NaOAc is included for comparison) [120] and electronegativity values [121]

| Compound | pK_b | Electronegativity of central Group V atom | | |
		Pauling	Allred-Rochow	Sanderson
NaOAc	6.58	–	–	–
Ph_3N	9.20	3.04	3.07	2.93
Ph_3P	8.57	2.19	2.06	2.16
Ph_3As	10.60	2.18	2.20	2.53
Ph_3Sb	8.76	2.05	1.82	2.19
Ph_3Bi	8.81	2.02	1.67	2.06

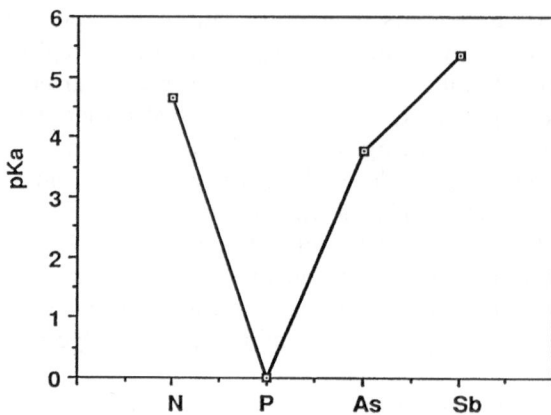

Fig. 19. Trends in pK$_a$ values of trimethyl Group V oxides in water. Data from Ref. [122]

In the case of the oxides of N, P, and As, the As and N compounds have been shown to be the more basic [122, 123]. For example, in the series Me$_3$EO where E is a Group V element, pK$_a$ measurements in H$_2$O show the order of basicity to be Sb > N > As > P, Fig. 19, and for the trioctyl compounds, (C$_8$H$_{17}$)$_3$EO, pK$_a$ values measured in nitromethane by potentiometric titration with perchloric acid showed that again the phosphine oxide is the least basic (pK$_a$ 9.72) with the amine and arsine compounds having a similar basicity (pK$_a$ 16.85 and 16.14 respectively) [123]. The pK$_a$ values of arsine oxides in this solvent are substantially lower (by a factor of $\approx 10^{12}$) than in H$_2$O. It has been noted [120] that the order of basicity of Group V oxide perchlorates in anhydrous acetic acid is As > N > P, i.e. solvation plays a major role. The basicity of these compounds in both nitromethane and water also depends on the nature of the substituent on the Group V atom and, as has been shown for As [123], on the increasing length of the carbon chain on the central atom. The pK$_a$ values of the related arsenyl sulphides are approx. 10^9 times less than those of the corresponding arsenyl oxides [123], thus the electron density at the oxygen atom is the dominant factor in the basicity of these compounds. Nevertheless the overall trends in basicity of these Group V oxides E(V) are the same as for the E(III) compounds, R$_3$E, i.e. E determines the electron density at O or S.

2.4.2 Metal Complexes

The compounds R$_3$E all have a lone pair associated with the central E atom and hence can act as Lewis bases. The order of basicity or electron-donating power of R$_3$E towards "soft" metal ion acceptors was proposed to be: N $\ll$ P > As > Sb > Bi [124]. This has been confirmed quantitatively for the metal ions Ag(I) [125, 126], Cu(I) [127], and Hg(II) [127] from stability constant measurements on metal-EPh$_3$ complexes as illustrated for Ag(I),

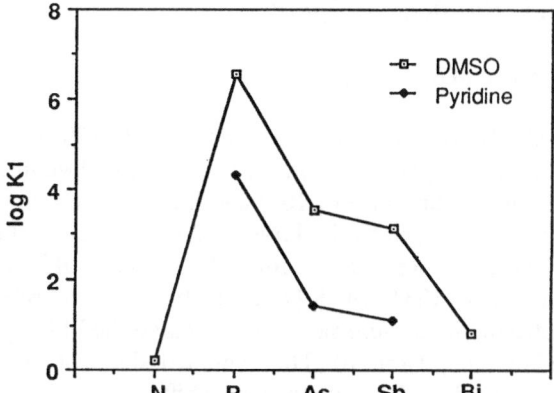

Fig. 20. Equilibrium constants for the binding of Ag(I) to Ph_3E in DMSO ($\square$) and pyridine ($\blacksquare$). Data from Ref. [126]

Fig. 20. The compounds were studied in DMSO [125, 127] and pyridine [126] because of the lack of aqueous solubility of Ph_3E. While the order of ligand affinity was the same in both solvents, complexes in pyridine were much less stable due to the stronger solvation of Ag(I) in pyridine, and only for PPh_3 were the three complexes $[Ag(PPh_3)_n]^+$ (n = 1 to 3) formed [126]. The greater affinity of P over As towards soft metal ions has also been shown in water [128] for Ag(I) complexes of the sulphonated ligands, $[m\text{-}(C_6H_4SO_3)Ph_2P]^-$ and $[m\text{-}(C_6H_4SO_3)_3As]^{3-}$. The affinities of the metals, Cu(I), Ag(I), and Hg(II), for a particular ligand (Ph_3E; E = P, As) increases in the order Cu(I) < Ag(I) < Hg(II) [127]. In the case of gold, both Au(III) and Au(I) have strong class (b) character, greater than that of Ag(I) [124]. Studies on the reduction of Au(III) by Ph_3E (where E is P, As or Sb) show that the reaction is quantitative in non-aqueous solution [129] and that the rate of reduction is greatest for Ph_3P. The equilibrium constants for the reaction:

$$[AuCl_2]^- + Ph_3E \rightleftharpoons AuCl(EPh_3) + Cl^-$$

showed the expected order of P $\gg$ As > Sb. Only for Ph_3P is the complex $[Au(Ph_3P)_2]^+$ formed. For the cytotoxic Au(I) complexes of bidentate Group V ligands (Sect. 3.3.3), ^{31}P NMR studies showed [97] that the bis ligand complex, $[Au(dadpe)_2]Cl$, (where dadpe is $Ph_2PCH_2CH_2AsPh_2$) is more labile in solution than the analogous phosphine complex, $[Au(dppe)_2]Cl$, (where dppe is $Ph_2PCH_2CH_2PPh_2$), and this difference is probably due to the weaker donating power of the Ph_2As group.

Gas-phase calorimetric studies of the reaction of Me_3E (E = P, As, Sb) with the borderline (i.e. soft/hard) boron trihalide acceptors again show [45] that Me_3P is the most strongly electron donating and Me_3Sb, the least, i.e. the order of Lewis basicity is P > As > Sb. This order in the gas phase contrasts with the proton acceptor ability of these centres as established by pK_a, pK_b measurements in solution.

2.5 Physical Methods

2.5.1 NMR

^{75}As NMR has not been widely used [130–132] because of its unfavourable nuclear properties, Table 9 [130, 133]. Although one stable isotope, ^{75}As, has a natural abundance of 100% with a nuclear spin of 3/2, its electric quadrupole moment is relatively large $(0.3 \times 10^{-24} \text{ cm}^2)$ [132], thus leading to broad resonances. Resonances for ^{75}As are detectable only in very symmetrical environments [133], e.g. K[AsF$_6$], the standard reference compound. A series of tetraalkylarsonium salts have been studied [132] with shifts between δ 206 and 258 p.p.m., Table 10. The chemical shifts of these compounds show high frequency β and δ substitution shifts and a low frequency γ effect. This is a reflection of the methyl substitution at positions β, γ, and δ to the As-position (α), e.g. substitution of a methyl group for a β–H to the As causes a deshielding of 10 to 11 p.p.m going from [AsEt$_4$]$^+$ to [AsPr$_3^i$Et]$^+$, while substitution at a γ

Table 9. NMR properties of Group V elements [130, 133]

Nucleus	Spin	Abundance (%)	Resonance Frequency (100 MHz ^{1}H)	Quadrupole Moment (10^{-28}m)	Receptivity/^{13}C
^{14}N	1	99.63	7.23	0.016	5.7
^{15}N	1/2	0.37	10.14	0	0.022
^{31}P	1/2	100	40.48	0	377
^{75}As	3/2	100	17.12	0.3	143
^{121}Sb	5/2	57.25	23.93	− 0.5	520
^{123}Sb	7/2	42.75	12.96	− 0.7	111
^{209}Bi	9/2	100	16.07	− 0.4	777

Table 10. ^{75}As NMR data for arsonium salts at concentrations of 0.1 M in water. Data are taken from Ref. [132], counterion [Br]$^-$, shift ref. KAsF$_6$, 0.1 M, 30°C, except where stated

Compound	δ (p.p.m.)	$\Delta v_{1/2}$ (Hz)
[AsPr$_3^i$Et]$^+$	258	750 (22°C)
[AsEt$_4$]$^+$	249	194
[AsEt$_3$Bu^n]$^+$	245	210
[AsEt$_3$Pr^n]$^+$	243	203
[AsEt$_3$Me]I	242	800
[AsBu$_4^n$]$^+$	234	346
[AsPr$_3^n$Et]$^+$	234	212
[AsPr$_3^n$Bu]$^+$	231	250
[AsPr$_4^n$]$^+$	230	230
[AsPr$_3^n$Me]I	225	Not given
[AsPh$_4$]Cl	217	1300
[AsMe$_4$]$^+$	206	113

position causes an upfield shift of 4–5 p.p.m. going from $[AsPr_4^i]^+$ to $[AsBu_4^n]^+$, Table 10. Such substituent effects are common in Group IV and V alkyls [130, 131]. The overall range of chemical shifts for ^{75}As so far observed [133] is 660 p.p.m, i.e. from 369 p.p.m. ($[AsO_4]^{3-}$) to -291 p.p.m. ($[AsH_4]^+$) though this covers only a few types of compounds and the actual range may be greater. This compares with ranges [130] of 720 p.p.m. for ^{31}P and 3450 p.p.m. for ^{121}Sb. Only two ^{209}Bi NMR spectra have been reported, those of $Bi(NO_3)_3$, with a linewidth of 3200 Hz, and $KBiF_6$ [133]. The ^{75}As NMR spectrum of a neutral As(V) compound, $AsCl_2F_3$, has recently been reported [134] (δ 205 p.p.m.; $\Delta v_{1/2}$ 5000 Hz at $-90\,^\circ C$). An ^{75}As NMR spectrum of the naturally occurring arsenocholine $\{[Me_3AsCH_2CH_2OH]Cl\}$, section 3.4, has been obtained [135], though attempts to acquire a spectrum of arsenobetaine were unsuccessful.

Only two types of coupling constants to ^{75}As have been measured. Coupling between ^{75}As and ^{19}F lies in the range 840 to 1048 Hz [133], e.g. for Me_3NAsF_5, $^1J(^{75}As-^{19}F)_{ax}$ 840 Hz, $^1J(^{75}As-^{19}F)_{eq}$ 1048 ± 8 Hz. For ^{75}As and ^1H, coupling constants from 92 to 555 Hz have been measured [133], e.g. for liquid AsH_3, $^1J(^{75}As-^1H)$ is 92.1 Hz; for $[AsH_4]^+$, $^1J(^{75}As-^1H)$ is 555 Hz. No coupling to ^{31}P, ^{13}C or to other ^{75}As nuclei has been observed.

Relaxation times of ^{75}As are generally very short (T_1 typically 10^{-2} to 10^{-4} s) [133] due to the efficient quadrupolar relaxation pathways available. They show the expected dependence on factors such as temperature, solvent, concentration. For example, the value of T_1(As) for AsH_3 varies from 48 µs at $-99\,^\circ C$ to approx. 95 µs at $25\,^\circ C$ [131].

^1H NMR studies on $(RAs)_n$ [65, 136, 137] and $(RAsO)_n$ [138, 139] have provided information on the solution structures of these compounds. The solution chemistry of liquid arsenomethane in particular has been studied by NMR [65] and it has predominantly a 5-membered ring in solution [136]. High temperature NMR spectra indicate arsenomethane is stable in solution to $200\,^\circ C$ [137]. The NMR spectrum of arsenobenzene has not been studied in detail [137]. The cyclic phosphorus compound $(PhP)_6$ exhibits solvent-dependent equilibria with $(PhP)_5$, e.g. in THF, CS_2, C_6H_6, the only species present is the hexamer, whereas in $CHCl_3$, the pentamer predominates [140]. When arsenomethane is treated with cacodyl, $Me_2As-AsMe_2$, an equilibrium mixture is obtained, involving the making and breaking of As–As bonds, the C–As bonds being stable [136]. The major species are $(MeAs)_5$ and cacodyl. ^1H NMR studies of arsenosomethane $(MeAsO)_n$ in solution suggest that, like arsenomethane, there is an equilibrium mixture with $n = 2$ to 5 and perhaps some linear polymer present also [138]. The predominant forms are tri- and tetra-meric [138, 139] with the O atoms in bridging positions [139]. Crystals of arsenosomethane are stable at room temperature for a long time, although after some months in solution, NMR results indicate that exchange of methyl groups occurs [138]. When arsenosomethane is treated with cacodylic oxide [138], $Me_2AsOAsMe_2$, an equilibrium mixture is obtained, containing compounds of the type $Me(AsMeO)_nAsMe_2$ ($n \geq 2$), i.e. effectively cacodylic oxide with

MeAsO groups inserted. The reported [141] ^{1}H NMR of arsenosobenzene in CDCl$_3$ ($\delta \approx$ 7.45 p.p.m., and 7.8 p.p.m. (both multiplets, relative intensity 2:1) does not help to assign the structure, however molecular mass measurements on this compound suggest that it too is cyclic in structure with n = 2 to 4, again with bridging O atoms [38] as is the cyclic structure determined for the S analogue, (C$_6$H$_5$AsS)$_4$ [44] (Sect. 2.2.1). Arsenosobenzene is commonly used as a titrant for direct and indirect measurement of residual chlorine and ozone in water [22]. However in view of the uncertain nature of the species, perhaps the results of such experiments should be treated with some caution.

It is of interest to note the relative trends of the ^{1}H NMR chemical shifts for the methyl groups of quaternary Group V complexes, Table 11 [53, 141, 142–152]. For the compounds [Me$_4$E]X in water (E = N [145], As[145], Sb [147]; X = I) there is an upfield shift in δ going from N to As to Sb (δ = 3.2, 1.91, 1.65 p.p.m., respectively), suggesting that the effect is dominated by the electronegativity (Pauling scale) of E with increasing electron withdrawal from and consequent deshielding of the methyl protons on going from Sb to As to N. The tetramethyl phosphonium salt follows this trend (δ = 2.47 p.p.m. [143]), but the reported spectrum was obtained in CDCl$_3$ so the shifts cannot be compared directly. Similar upfield trends have been observed for the naturally-occurring arsenobetaine [53] and betaine [148] (δ = 1.87 and 3.27 p.p.m. respectively), and for choline [148], acetylcholine [151], phosphorylcholine [148] and the arseno-analogues, aresenocholine [150], acetylarsenocholine [150] and phosphorylarsenocholine [150]. In the case of phosphocholine and phosphorylphosphocholine [149], the chemical shift is to higher field of those of the N or As derivatives. This has also been noted for the shifts of the methyl group in the Me$_3$EO series (δ = 3.32, 1.56, 1.69 p.p.m. for E = N, P, As respectively) [122]. The reasons for these trends are not clear. For the complexes [Ph$_3$EMe]X (E = P, As) with the counterion X = BPh$_4$, the values for E = P and E = As are similar [142] (δ = 1.915, 1.935 p.p.m. respectively) and in this case the shift is influenced by the shielding effect of the Ph groups of the anion. The reduction of the shielding effect of the counter ion on the methyl group in the Sb case was attributed to the greater electronegativity of Sb relative to As, P [142]. It is interesting to note that the linewidths of peaks for [NEt$_4$]$^+$ and [NMe$_4$]$^+$ are independent of solvent, whilst those of [AsEt$_4$]$^+$, [AsMe$_4$]$^+$, [SbEt$_4$]$^+$ show pronounced effects [145].

2.5.2 Other Spectroscopic Methods

Other spectroscopic methods useful in studying As compounds include nuclear quadrupole resonance and Mössbauer spectroscopies.

The theory and applications of nuclear quadrupole resonance (NQR) have been reviewed by Buslaev et al. [153]. The nature of the spectra obtained is related to the electron distribution around the nucleus in question and thus they give information on the structure and bonding in complexes. NQR studies have

Table 11. ^{1}H NMR chemical shifts of methyl groups (and CH_2 groups where stated) bonded to quaternary Group V atoms

Compound	δ (p.p.m.)			
	E = N	E = P	E = As	E = Sb
[Ph$_3$EMe]X [142][1]				
X = BF$_4$ [142]	–	–	2.785	2.500
X = (BF$_4$)(BPh$_4$) [142]	–	–	–	2.1150
X = BPh$_4$ [142]	–	1.915	1.935	1.785
X = I [143]	–	3.12	–	–
[Ph$_2$MeEMe]I [144][2]	–	2.89	–	–
[PhMe$_2$EMe]I [144][2]	–	2.67	–	–
[EtMe$_2$EMe]I [145]	3.07[3]			
[Me$_3$EH]Cl [143][1]	–	2.30		
[Me$_4$E]X			1.7 [146][4]	
X = Cl	3.2 [141]	–	–	–
X = I	3.20 [145][3]	2.47 [143][1]	1.87–1.95 [145][3]	1.65 [147][5]
[Me$_3$ECH$_2$COO]	3.27 [148][6]	–	1.87 [53][7]	
	3.91 (CH$_2$)		3.28 (CH$_2$)	
[Me$_2$ECH$_2$CH$_2$OH]Cl	3.21 [148][8]	1.8 [149][7]	2.44 [150][9]	–
(Choline)	4.1 (CH$_2$)	3.7–4.2[10] (CH$_2$)	4.47 (br) (CH$_2$)	
	3.5 (CH$_2$)	2.28–2.7[10] (CH$_2$)	3.13(t)(CH$_2$As)	
[Me$_3$ECH$_2$CH$_2$OCOCH$_3$]Br	3.23 [151][11]	–	2.50 [150][9]	
(acetylcholine)			5.00(t)(CH$_2$)	
			3.32(t)(CH$_2$As)	
[Me$_3$ECH$_2$CH$_2$OPO$_3$H$_2$]X	3.23 [148][12]			
(phosphoryl choline) X = Cl	–	2.01 [149][7]		
		4.53–3.9 (CH$_2$)(pr of q)		
		2.81–2.4 (CH$_2$)(pr of t)		
X = Br	–	–	2.45 [150][9]	–
			4.85(CH$_2$)(d of t)	
			3.30(CH$_2$)(t)	
[Ph$_2$MeECH$_2$EPh$_2$]I	–	2.77 [152][13]	–	–
[Ph$_2$MeECH$_2$EMePh$_2$]				
[SO$_3$Fe]$_2$	–	2.25 [152][14]		

[1] For all except X = I, solvent: CD_2Cl_2 at 60 MHz; for X = I, solvent: $CDCl_3$ also at 60 MHz
[2] Solvent: $CDCl_3$; 100 MHz
[3] Solvent: H_2O; 60 MHz; 35 °C; for [Me$_4$As]I shift is concentration dependent.
[4] X unknown, may be Cl$^-$ or OH$^-$; isolated from the clam, *Meretrix lusoria*; solvent: D_2O, 100 MHz
[5] Solvent: H_2O; 60 MHz; 35 °C
[6] In D_2O at pH* 6.7 at 500 MHz
[7] Solvent: D_2O at 60 MHz; pH not given
[8] In D_2O at pH* ~6 at 500 MHz
[9] In D_2O; 100 MHz with external TMS as reference. No pH given.
[10] Pair of triplets; the more upfield signals are an overlapping pair. pH of solution not given
[11] Counterion [I]$^-$; 90 MHz in D_2O
[12] Run as glycerol phosphoryl choline in D_2O, pH* 6.7
[13] In $CDCl_3$. Frequency not given
[14] In CF_3COOH. Frequency not given

been reported for the Group V isotopes ^{75}As, ^{121}Sb, ^{123}Sb, and ^{209}Bi. NQR studies of arsenic compounds have been reviewed by Zakirov et al. [154]. The arsenates R_2HAsO_4 and RH_2AsO_4 have been extensively studied by ^{75}As NQR because of their ferroelectric and antiferroelectric properties [153]. The

^{75}As NQR frequency of the tetrahedral ion $[AsO_4]^{2-}$ is dependent on the degree of distortion, which in turn affects the polarisation of the crystals. A series of donor–acceptor complexes, $EX_3.Ar$, $EX_3.2Ar$ (Menshutkin complexes, where E is As, Sb, Bi; X is Cl, Br; Ar is an aromatic molecule) have been studied by NQR [153]. In several of the $AsX_3.Ar$ complexes, the ^{75}As NQR frequency decreases considerably with respect to pure AsX_3, the ^{75}As frequency shifts showing a linear correlation with the ionization potential of the aromatic molecule. The structures of some Sb and Bi halides and oxyhalides and of As, Sb and Bi chalcogenides have also been studied by NQR and the data compared with X-ray results where available, e.g. As_2S_3. Other As compounds studied by NQR include the cyclopolyarsines, $(PhAs)_6$ and $(C_6F_5As)_4$ [65, 155], trivalent arsenites, $As(YR)_3$, (Y is O, S; R is alkyl or aryl) [119], and $Ph_2As-R-AsPh_2$ (where R is CH_2, $CH_2C_6H_4CH_2$, $(CH_2)_4$, $(CH_2)_{10}$) [156]. From the NQR data obtained for $Ph_2As-R-AsPh_2$ and analogous Sb compounds (and Mössbauer spectra of the latter) it was concluded that the As or Sb–methylene bond is electronically similar to the As or Sb–phenyl bond [156].

Arsenic itself is Mössbauer-silent; however some studies have been done on Fe compounds with Group V ligands, and from the quadrupole splittings the relative π-acceptor strengths of the ligands can be seen, e.g. in $LFe(CO)(NO)_2$ the order of π-acceptor ability is $L = PPh_3 < AsPh_3$ [157]. Also, in $Fe(CO)_3L_2$, $(L = PPh_3, AsPh_3, SbPh_3)$, the P compound shows the smaller Q.S. [158], while those of the As and Sb compounds are similar and it was concluded that P is the stronger σ donor, as reported previously [159] for $LFe(CO)_4$ and $L_2Fe(CO)_3$ where L is EPh_3 (E = P, As, Sb), Table 12. In general the isomer shifts for ^{197}Au in gold(I) phosphine complexes are greater than those in the analogous arsine complexes (Table 12), consistent with phosphines being the stronger σ donors and stronger π acceptors [160].

The infra-red and Raman vibrations of EH_3, (E = P, As, Sb), $[EH_4]^+$ and $[EO_4]^{3-}$ (E = P, As) have been assigned, Table 13 [161]. The stretching frequencies for the Group V-oxide, E=O bonds are 1195 and 880 cm^{-1} for E = P and E = As, respectively [161]. Infrared spectroscopy shows the absence of As=O in the arsenoso compounds [162]. The i.r. spectrum of arsenosoben-zene shows a broad and intense absorption in the 715 to 750 cm^{-1} region [163] similar to that of $(Ph_2As)_2O$ [38] and a weak absorption at 910 cm^{-1} too weak to be assigned to As=O. This suggests that the species is cyclic or polymeric, containing As–O–As bridges as suggested by molecular mass measurements, Sect. 2.5.1. Raman spectroscopy has been used to establish the solution structure of arsenious acid (Sect. 2.3).

2.5.3 Other Methods

The most common method of determining the total arsenic content of a sample is atomic absorption spectrometry [164] via generation of volatile hydrides. The usual reductant is sodium borohydride [164]. The reduction is pH dependent,

Table 12. Mössbauer parameters for some complexes of Group V ligands

Compound	Nucleus	σ (mm s^{-1})	Δ (mm s^{-1})	Ref.
L$_2$Fe(CO)$_3$	^{57}Fe[1]			
L = PPh$_3$		0.16	2.63	[158]
L = AsPh$_3$		0.20	3.20	
L = SbPh$_3$		0.19	3.16	
L$_2$Fe(CO)$_3$	^{57}Fe[1]			
L = PPh$_3$		0.06	2.62	[159]
L = AsPh$_3$		0.16	3.24	
L = SbPh$_3$		0.19	3.16	
LFe(CO)$_4$	^{57}Fe[1]			
L = PPh$_3$		0.07	2.40	[159]
L = AsPh$_3$		0.10	2.71	
L = SbPh$_3$		0.10	2.67	
(SbPh$_3$)$_2$Fe(CO)$_3$	^{121}Sb[2]	1.5	18	[159]
(SbPh$_3$)Fe(CO)$_4$	^{121}Sb[2]	2.8	16	[159]
L$_2$Fe(NO)$_2$	^{57}Fe[1]			
L = PPh$_3$		0.33	0.70	[157]
L = AsPh$_3$		0.42	0.59	
LFe(NO)$_2$Br	^{57}Fe[1]			
L = PPh$_3$		0.50	1.02	[157]
L = AsPh$_3$		0.20	1.28	
L = SbPh$_3$		0.55	1.49	
[Au$_2$Ru$_4$(μ_3-H)(μ-H)L$_2$(CO)$_{12}$]	^{197}Au[3]			
L$_2$ = Ph$_2$PCH$_2$PPh$_2$		3.02	6.71	[160]
L$_2$ = Ph$_2$AsCH$_2$PPh$_2$		2.9[4]	7.1	
		2.5[5]	5.5	
L$_2$ = Ph$_2$AsCH$_2$AsPh$_2$		2.53	5.93	
L$_2$ = Ph$_2$PCH$_2$CH$_2$PPh$_2$		3.1	7.2	
		2.5	5.8	
L$_2$ = Ph$_2$AsCH$_2$CH$_2$PPh$_2$		2.6[4]	7.0	
		2.2[5]	5.3	
L$_2$ = Ph$_2$AsCH$_2$CH$_2$AsPh$_2$		2.4	6.6	
		2.3	5.1	
[AuClL]	^{197}Au			
L = PPh$_3$		4.08	7.43	[160]
L = PMe$_2$Ph		3.91	7.11	
L = AsPh$_3$		3.08	6.96	
L = AsMe$_2$Ph		2.93	6.37	

[1] Run at 80 K and isomer shifts quoted w.r.t. sodium nitroprusside unless otherwise stated;
[2] 80 K, isomer shifts quoted w.r.t. InSb; values for SbPh$_3$ are -0.9 ± 0.2 and 17 ± 2 for I.S. and Q.S. respectively;
[3] At liq. He temperature;
[4] Au-P;
[5] Au-As.

thus it is sometimes possible to distinguish between As(III) and As(V). For organoarsenic compounds not reduced to volatile hydrides, e.g. arsenobetaine and arsenocholine, a convenient method of analysis involves initial conversion of the arsenic compound to arsenate via photooxidation [165], prior to determination of the arsenate by atomic absorption. Other instrumental methods used include neutron activation analysis [164], which involves irradiation of the sample with γ-rays converting ^{75}As to ^{76}As. This again is a non-specific method

Table 13. Vibrational frequencies (cm^{-1}) for some Group V compounds [161]

Compound	v_1	v_2	v_3	v_4	Comment
EH_3					
$\quad E = P$	2327	991(d)	2421	1121	Gas phase; all i.r. and Raman active
$\quad E = As$	2122	906	2185	1005	
$\quad E = Sb$	1891	782	1894	831	
$[EH_4]^+$					
$\quad E = P$	2304	1040	2370	930	Only v_3 and v_4 i.r. active
$\quad E = As$	2119	1002	2225	845	
$[EO_4]^{3-}$					
$\quad E = P$	938	420	1017	567	All i.r. and Raman active
$\quad E = As$	837	349	878	463	

and may be slow especially with biological samples since one of the interfering agents is ^{24}Na ($t_{1/2} = 15.3$ h). Other less common methods for As determination have been reviewed by Brooks et al (to 1980) [164].

Speciation of mixtures of arsenic compounds is usually attempted by chromatographic means. For example, $CH_3As(O)(OH)_2$, $(CH_3)_2As(O)(OH)$, and inorganic As(III) and As(V) have been determined in water and urine by GLC using electrochemical detection [166]. Separation has also been effected with ion and high pressure liquid chromatographies [167] followed by determination with element-specific detectors [168], e.g. inorganic As(III) and As(V) together with $CH_3As(O)(OH)_2$ and $(CH_3)_2As(O)(OH)$ have been separated on a cation–anion exchange column bed and detected by flameless atomic absorption spectrometry with a limit of detection of <1 µg/sample [169]. An even more sensitive method of detection is ICP (inductively coupled argon plasma emission spectrometry) coupled to HPLC as separating method [168, 169]. Separation of a mixture of arsenate, arsenite, $CH_3As(O)(OH)_2$, $(CH_3)_2As(O)(OH)$, and arsenobetaine with detection in the 20–40 ng range by HPLC-ICP has been achieved [169].

3 Biological Activity

3.1 Metabolism of Arsenic Compounds

3.1.1 Essentiality of Arsenic

An element may be described as essential for a particular organism, if its absence prevents the completion of the life cycle [170], i.e. if it affects the growth, reproduction, or life expectancy [171]. In addition, the deficiency symptoms of an essential element may be alleviated only by the direct action of the element on the nutritional status of the organism [170], and will reoccur on removal of the element from the diet [171]. An essential element should have a known metabolic role. For micro-nutrients, i.e. those elements for which the dietary requirement is very low, experimental problems arise in eliminating naturally occurring traces of the substance from feedstuffs. This was the case with the early work on As essentiality [171] since levels of $< 50 \, \mu g/kg$ As/dry matter are necessary before deficiency symptoms are observed. Furthermore, the metabolic biochemistries of many elements are interdependent; deficiency in one may exacerbate the effect of deficiency of another, e.g. As and Zn, see below [172, 173]. An additional problem with these types of definition is that they neglect the differences between different forms of the element. For example, As(III) and As(V) have different mechanisms of toxicity, while many organoarsenic compounds are non-toxic and are not metabolised by man (Sect. 3.4). A further point of importance is that rats metabolise arsenic differently to other animals in that they appear to store it in red blood cells associated with the haemoglobin or haematin [174]. It has been suggested that decreased methylation of As is the reason for the different metabolism [174]. Thus investigations using rats cannot be generalised [174, 175].

It has been claimed [2, 176] that peasants in the Austrian province of Styria, and in the Alps of Switzerland and the Tyrol, used to eat quantities of an arsenic compound, variously described as blackish-grey elemental As [2], or white arsenic trioxide [177] in order to increase their weight, strength, appetite, to aid endurance at high altitudes, and to clear the complexion. This has been dismissed as a legend by Lenihan [177], but others have described first hand accounts of the practice [135].

Organoarsenic compounds have been used as growth promoters in pigs and poultry, Sect. 3.3.1. Symptoms such as reduction in growth (of goats) and fertility and viability of offspring (of goats and mini-pigs) have been observed in animals on As-deficient diets ($< 50 \, \mu g/kg$) [171]. Histological changes in the cardiac muscle and liver of As-deficient goats were observed suggesting that there had been an effect on the mitochondria [171]. In chicks it was found that the levels of As (administered as Na_2HAsO_4), Zn, and arginine were interdependent [172, 173] together influencing the nutritional status of the animal. Though the mechanism of interaction is unknown, it has been postulated that As plays a role in amino acid/protein metabolism possibly via an alteration of arginine metabolism [173]. Arginine is now known to play a vital role in the production of the muscle relaxant nitric oxide in vivo [178] and arsenic could interact with some of the metalloenzymes (Cu, Fe) in this pathway. Although the animal studies suggest that As is essential, to date no As-dependent metabolic pathway has been proven, thus according to the criteria listed above it is not essential.

No primary signs of As deficiency in animals are apparent [179]. Nevertheless, there are warnings of the possibility of future deficiencies in dietary arsenic [180], and the possible consequences of a non-specific ban on the use of arsenicals in animals destined for human consumption introduced in the FRG in 1981 have been outlined [181].

3.1.2 Methylation

Methylation is the most important route by which As(III) is metabolised in man and most other animals. Many reviews on the methylation of arsenic and other elements have been published [182, 183] and methylation in both mammals [8, 184, 185] and non-mammals [12, 184, 185] has been discussed.

The methylation of arsenic compounds by several species of fungi and bacteria has been extensively studied [12] since the pioneering work of Challenger [186] who proposed that methylation is an oxidative-addition process, Fig. 21, with addition of $[CH_3]^+$ from a donor such as S-adenosyl-methionine (SAM). Subsequent work [12] has confirmed that all the proposed As(V) intermediates are present in cell preparations from $C.$ $humolica$ and that SAM is the probable methyl donor. However the As(III) species in brackets in Fig. 21 are not known. Both methylarsonous acids, $MeAs(OH)_2$, and dimethyl-arsinous acid, Me_2AsOH, are probably better formulated as hydrated $(MeAsO)_n$ and $(Me_2As)_2O$ species, Sect. 2.2.1. The metabolism of $(MeAsO)_n$ to $Me_2As(O)OH$ has been observed in $C.$ $humolica$ [187]. Reduction of Me_3AsO to Me_3As by a variety of microorganisms has been demonstrated [188] including $C.$ $humolica$ and a marine pseudomonad. Methylcobalamin, while not the primary methyl donor for methanogenic bacteria, can act as such in cell extracts [12]. Model studies on the reaction of methylcobaloxime [methyl(pyridinato)bis(dimethylglyoximato)Co(III)] and methyldihaloarsines in $CHCl_3$ using NMR suggest [184, 189] that the mechanism involves initial coordination of the arsine to the Co, migration of the Me group to an equatorial position in the cobaloxime followed by Me transfer of this group to the As. However, other workers failed to observe any reaction between this meth-ylcobaloxime and $MeAsCl_2$ [82].

Methylation of arsenic by fungi has caused problems in the past. As(III) derivatives {formulated as $[Cu_2(C_2H_3O_2)(AsO_3)]$, Paris Green, and $[CuHAsO_3]$, Scheele's green [36]}, were widely used in the late eighteenth and early nineteenth centuries as pigments in paints and wallpapers [37]. Methyl-ation of these pigments by moulds such as $S.$ $brevicaulis$ growing on damp wallpaper afforded volatile Me_3As [37]. There are a number of reports of people being made ill or even killed by their wallpaper and this phenomenon has been implicated in the possible arsenic poisoning of Napoleon Bonaparte [37], though other workers have failed to find abnormal levels of As in samples of his hair [190]. Methylation also plays an important role in the metabolism of arsenic in the marine environment and this is discussed in Sect. 3.4.

Studies on the mechanism of methylation of As in mammals including man are reviewed in Refs [182, 185]. When As(III) (as sodium arsenite) is adminis-tered to humans intravenously, it is found that As is excreted in the urine over

24 h as dimethylarsinate (24%), methylarsonate (13%), as well as inorganic As(III) (66%) and that liver disease increases the proportion of dimethyl-arsinate, and decreases the proportion of methylarsonate excreted [185]. From this it has been concluded that there are two different pathways for the production of mono- and di-methyl species. Man is the only mammal to excrete significant amounts of monomethyl arsonic acid [182] with approx. 25% of the total As excreted in this form. It can also be assumed that the site of methylation is the liver [182] since in vitro studies have shown [174] that no methylation of As occurs in the urine, plasma, or red blood cells of man or dog. It has been postulated that intestinal bacteria may play a role in the process [8], however germ-free mice have been shown to methylate As to the same extent as ordinary mice [174], thus intestinal flora are not the major source of methylation. Thus while much of the arsenite is excreted unchanged, some is methylated. Studies on rates of renal excretion of arsenite, and the mono and dimethylated species in man have shown [175] that the methylated species are excreted more rapidly, methylarsonate being excreted most rapidly. The biological half-life of arsenite is about 30 h in humans [175]. Studies in mice suggest that methyltransferase enzymes and SAM are the methyl donors [185]. In vivo [191] and in vitro [192] studies on methylation of As(III) in rats suggest that GSH plays an important role in the process, perhaps by protection of labile thiol groups, activation of methyltransferase enzymes, or by preventing the oxidation of As(III) [192]. From the in vitro work [192], SAM is the methyl donor. However as noted previously rats metabolise As in a different way to other mammals [174, 175], so these observations should be interpreted with caution. Depletion of liver GSH by diethyl maleate prior to orally administering arsenite to mice does not affect the excretion of dimethylarsinic acid [182]. One mammal is known which appears unable to methylate As, the Marmoset monkey [185], and it has been suggested that this is because of its binding to microsomes in the liver and retention in red cells. Since this species can survive without methylation it has been postulated by Aposhian [185] that protein binding rather than methyl-ation is the primary mechanism of mammalian detoxification of As.

As(V) as arsenate is not directly methylated [184] and so must first be reduced to As(III). Model studies on the reactions of cacodylic acid [193] and of trimethyl arsine oxide [194] with thiols suggest that this is a likely method of reduction of the As(V) affording As(III) thiol containing intermediates. These products decompose in the presence of oxygen giving disulphides and the As(V) acid, thus such intermediates may not readily be seen in vivo. Results from spin-echo NMR experiments on the interaction of cacodylic acid with erythrocytes [12, 195] suggest that the As(V) acid is reduced by glutathione to an $Me_2As(III)$-S species possibly bound to a transmembrane protein [12]. Cullen et al. [193, 194] have isolated and characterised by elemental analysis and ^1H and ^{13}C NMR, products from reactions of dimethylarsinic (cacodylic) acid, disodium methylarsonic acid and trimethylarsine oxide with a variety of thiols such as cysteine, glutathione, mercaptoethanol and lipoic acid. The reactions involve reduction of As(V) followed by formation of an S-dimethylarsino

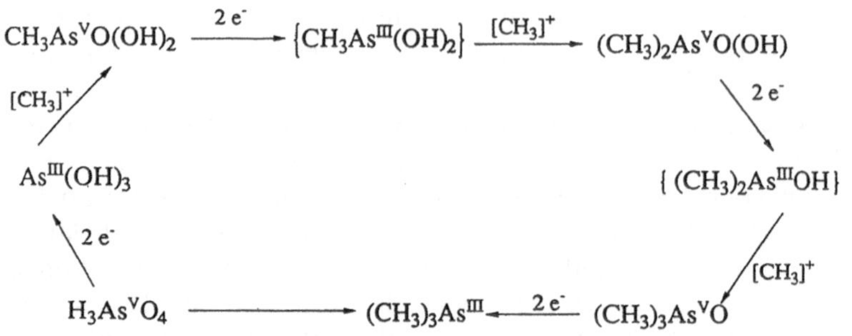

$$CH_3As^VO(OH)_2 \xrightarrow{2\,e^-} \{CH_3As^{III}(OH)_2\} \xrightarrow{[CH_3]^+} (CH_3)_2As^VO(OH)$$

with:

- $[CH_3]^+$ branch down-left to $As^{III}(OH)_3$
- $As^{III}(OH)_3 \xrightarrow{2\,e^-}$ down to $H_3As^VO_4$
- $(CH_3)_2As^VO(OH) \xrightarrow{2\,e^-}$ down-right to $\{(CH_3)_2As^{III}OH\}$
- $\{(CH_3)_2As^{III}OH\} \xrightarrow{[CH_3]^+}$ down to $(CH_3)_3As^VO$

$$H_3As^VO_4 \longrightarrow (CH_3)_3As^{III} \xleftarrow{2\,e^-} (CH_3)_3As^VO$$

Fig. 21. Proposed methylation pathway for arsenic metabolism. Adapted from Ref. [12]

compound. For example, for cacodylic acid:

$$Me_2As(V)O(OH) + 2RSH \rightarrow Me_2As(V)(SR)_2(OH) + H_2O$$

$$\rightarrow Me_2As(III)(OH) + RSSR$$

$$Me_2As(III)(OH) + RSH \rightarrow Me_2As(III)(SR) + H_2O$$

In general the S-dimethylarsino derivatives are unstable in the presence of O_2 and decompose to give disulphides and As(V).

Organoarsenic compounds present in crabmeat are excreted unchanged by man [175], however a later study on the ingestion of prawns containing an undefined organoarsenic compound showed that while 90% of it was excreted unchanged, approx. 3–5% was excreted as inorganic arsenic, dimethylarsinate and monomethylarsonate suggesting that there may be a demethylation pathway in man [185]. Dimethylarsinate is the major metabolite of inorganic arsenic in mammals; when it was administered orally to mice and humans, about 80% and 60% was excreted as unchanged dimethylarsinate and only 5% and 4% as trimethylarsine oxide in mice and man respectively [185]. Reduction of arsenate to arsenite before methylation has been shown to occur in rats and rabbits [182] from labelling studies with ^{74}As-arsenate. The main site of reduction is probably in the blood. Such reductions also occur in man since ingestion of a seaweed sample containing 86% of the arsenic as As(V) resulted in the excretion of 36% As(III) within 48 h [182]. Thus methylation in man probably follows the scheme described above (Fig. 21) without the formation of trimethylarsine. There appear to be no reports of the methylation of P in vivo, probably a consequence of the difficulty of reducing P(V) to P(III).

3.1.3 Enzyme Inhibition

One of the major causes of arsenic(III) toxicity is thought to be its ability to inhibit a variety of enzymes. The usual mode of action is often stated to be with thiol (Cys) groups, though with cholinesterase [196] the interaction must be different as it is not a thiol-containing enzyme. As discussed (Sect. 3.1.2), thiol groups may also reduce As(V) to As(III) prior to methylation.

Studies on the interaction of As(III) with thiols have been carried out since the early part of the century as described by Squibb and Fowler [5] and Webb [197]. Much of the earlier work on these interactions was prompted by the use of arsenicals such as *trans*-ClCH=CH-AsCl$_2$, Lewisite, as chemical warfare agents [198]. It was found that the inhibitory action of Lewisite on enzymes was reversible on treatment with a dithiol antidote, 2,3-dimercaptopropanol (British Anti-Lewisite or BAL) [5]. This, together with the fact that most of the enzymes shown to be inhibited by As(III) in vitro are thiol-containing enzymes or have thiol cofactors, e.g. lipoate [5, 197], led to the conclusion that the inhibition was due to the formation of an As–S enzyme complex. Such enzymes include [5] dehydrogenases (e.g. β-hydroxybutyrate dehydrogenase, pyruvate dehydrogenase, lipoic acid dehydrogenase), oxidases (e.g. cytochrome oxidase), and urease. Arsenite is, in general, less inhibitory than the organo As(III) compounds [2]. From the work of Peters in the 1940s and early 1950s, it appears that the α-keto oxidases, e.g. pyruvate oxidase, are specifically inhibited by As(III) [197] and it has been assumed that the mechanism of action involves formation of a cyclic As(III)-dithiol with the lipoate coenzyme, Fig. 22. It has been proposed from these results and others that monosubstituted As(III) compounds, e.g. (RAsO)$_n$ react with enzymes containing two adjacent thiol groups forming a ring, while compounds of the type R$_2$AsOH, i.e. disubstituted As(III) compounds, react only with a single SH group. These proposals have been used [5, 197] to classify enzymes as mono- or di-thiol containing, however the validity of the conclusions based on these has been questioned by Webb [197]. For example, inhibition of an enzyme by R$_2$AsCl does not imply a monothiol enzyme since the As compound could equally react with one thiol of two adjacent SH-groups, and lack of inhibition by (RAsO)$_n$ does not necessarily mean there are no dithiols; it could mean that the binding is weakened by other factors.

There is no direct conclusive evidence for the details of the mechanism of the interaction of As(III) with thiol-containing enzymes. Arsenious acid [As(OH)$_3$ at pH 7] could bind and disrupt H-bonding networks in proteins, bind as an arsenite anion, or act as a reducing agent [197]. The inhibition by organo-As(III) compounds may involve binding to sites other than the active site and

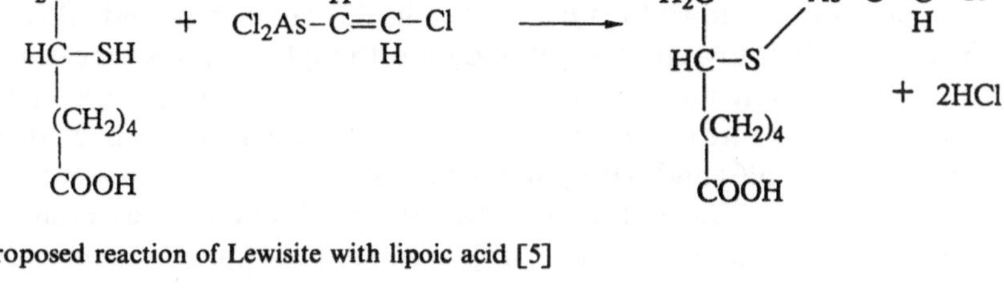

Fig. 22. Proposed reaction of Lewisite with lipoic acid [5]

induce conformational changes at the active site [197]. It is remarkable that no structural studies on As–S enzyme complexes or model compounds appear to have been reported, and it has been noted [185] that no studies on the biotransformation of As with *purified* mammalian enzymes exist in the literature. Recent studies with ^{14}C labelled phenyl dichloroarsine have shown that the arsenic compound (probably as the phenylarsonous acid) enters human erythrocytes in vitro and becomes associated almost entirely with the GSH fraction. The ratio of arsenic compound bound in the cell to GSH is approximately 1:2. NMR studies on the independently prepared complex, $PhAs(GS)_2$, suggest that there are two stereoisomers [199].

It should also be noted that the nature of many of the As compounds used in reported enzyme inhibition studies is not clear, e.g. phenyl arsine oxide (arsenosobenzene) used in β-hydroxybutyrate dehydrogenase inhibition does not contain As=O bonds but may be cyclic or polymeric. Also the compounds $R_2As(OH)$ may be hydrates of arsenic oxides $(R_2As)_2O \cdot nH_2O$ [38] (Sect. 2.3). Furthermore at pH 7 arsenite should be described as arsenious acid (pK_1 9).

In vitro studies on the inhibition of acetylcholinesterase (from the electric eel) and butyrylcholinesterase (human serum) by arsenious acid [196] led to the suggestion that two tyrosine residues on the enzyme were involved since 12 of the 21 tyrosine residues on the free enzyme were nitratable, compared with only 10 on the inhibited enzymes; however, this could well be an indirect effect. Studies on mice brain slices in vitro showed [200] that sodium arsenite had a stimulatory effect on the activity of the thiol-containing-enzyme, choline acetyltransferase, in contrast to other thiol-alkylating agents. Phenyl arsine oxide (arsenosobenzene) has been shown to interrupt signal transmission from the insulin receptor to the glucose transport system in 3T3-L1 adipocytes [201] perhaps by complexation to a cellular dithiol in the signal transmission pathway, though again there is no structural evidence of this mechanism, and it has also been shown to inhibit hepatic autacoid-stimulated vasoconstriction and glycogenolysis [202]. The biological activity of synthetic $R_2As(III)$-S compounds is discussed in more detail in Sect. 3.3.3.

3.2 Arsenic and the Immune System

Amino acid and protein conjugates of the compound, *p*-azophenylarsonate (*p*-$N_2C_6H_4AsO(OH)(ONa)$] (APA) affect the immune system. Conjugation occurs via binding of the azo-group binding to the NH_2 groups of the amino acids on the proteins as shown for tyrosine in Fig. 23. The compound APA is itself a hapten, i.e. it binds to a specific antibody site, though on its own it cannot induce antibody formation unless conjugated to a carrier. Initial work in this field focussed on the nature of the antibody binding site by use of chemically modified antibodies [203–206] and APA conjugated to unspecified components of bovine serum [203], or bovine γ-globulin [204, 205, 206]. Evidence was found for the presence of tyrosine or cysteine [203], arginine [204], guanidinium [205] and

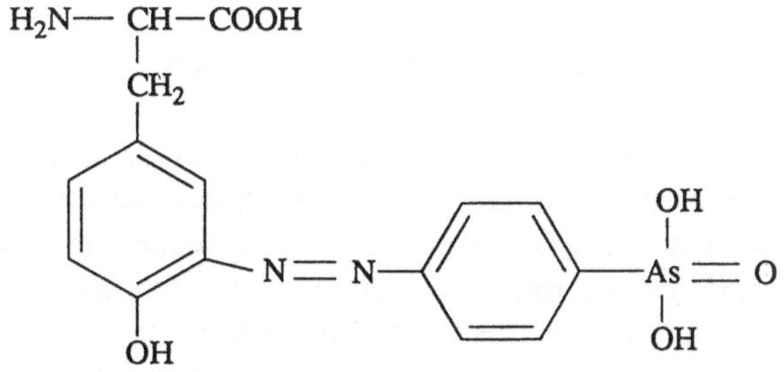

$$H_2N—CH—COOH$$

Fig. 23. Proposed structure of 4-azophenylarsonate tyrosine conjugate

ammonium [205, 206] groups in the binding sites. The interaction appeared to be predominantly electrostatic, though other factors are involved also, e.g. the antibodies induced by APA-bovine serum are not bound by azobenzoate [203]. Subsequently the N-terminal V region (the antigen-binding region) of the antibody produced when A/J mice were immunised with APA-keyhole limpet haemocyanin was sequenced [207] and it was found that there were three distinct antibody families, Ars A, Ars B, and Ars C. The molecular genetics of this system have been studied [208] as a model for more complicated immune response mechanisms.

Other work with APA has centred on the fact that when conjugated to tyrosine or N-chloroacetyl-tyrosine it induces delayed hypersensitivity (DH), a response associated with the T cells. It was found that substitution of arsonate on the phenyl-azo-tyrosine backbone by other charged groups such as sulphonate or trimethylammonium did not affect immunogenicity, though specificity was altered [209]. It has been suggested that the hydrophobic core of APA-tyrosine, i.e. the phenyl rings, is important in interaction with MHC (major histocompatibility complex) molecules which are responsible for immunogenic effects while the hydrophilic portion, including the arsonate group, is of importance in antibody binding [209]. It does not induce antibodies, thus it is a useful immunogen for studying DH effects. This area has been reviewed by Bullock [210] and later studies reported by Alkan [211]. A further aspect of this work is the ability of APA-tyrosine when linked to a chemical group acting as a hapten such as dinitrophenyl, DNP, to act as a carrier [212]. In this case a DH reaction due to the APA was observed together with antibody production for the DNP. Work on this in relation to lymphocyte activation has been reviewed by Goodman et al. [212]. In azophenylarsonate and its protein conjugates, As is in the $+5$ oxidation state and all the reported work assumes that the species is unchanged in vivo. It has been shown [193] that dimethylarsinic acid, $Me_2AsO(OH)$, can react with a variety of mono- and di-thiols including cysteine in vitro giving As(III) species, $Me_2As(SR)$ and earlier work on the reaction of aromatic As(V) acids such as p-$(NH_2)C_6H_4AsO(ONa)_2$ with thiols, for example, $HSCH_2CO_2H$ [213] has also been reported, the product being a thioarsinite, p-$(NH_2)C_6H_4As(SCH_2CO_2H)_2$. As discussed in Sect. 3.1.3, ^1H NMR studies of human erythrocytes have shown [195] that dimethylarsinic acid is reduced by glutathione to an $Me_2As(III)$-S species possibly bound to a transmembrane protein [12]. Such reactions may be important when APA interacts with the Ars A, Ars B, or Ars C antibodies if they contain free Cys groups.

It has been found [214] that sodium arsenite, As(III), has an immunosuppressive effect on male mice at low concentrations. It has previously been shown [5] that at low concentrations arsenite enhances the antiviral activity of interferon in mice, though at high concentrations inhibition of interferon production was observed. Moderately high doses of sodium arsenate, As(V), increase the mortality of mice exposed to virus-inducing agents, while arsanilic acid, an organo As(V) species, has no effect.

3.3 Antimicrobial, Anticancer and Antiviral Activity

This section considers the activity of arsenic compounds as antimicrobial/antiparasitical agents, and as anticancer and antiviral agents. Mechanisms of resistance of bacteria to certain arsenic compounds are also described, and, where appropriate, the activity of other group V compounds is mentioned.

3.3.1 Antimicrobial Activity

The most widespread use of arsenic compounds in therapeutics has been in the treatment of syphilis, trypanosomiasis and amoebic dysentery [3, 197]. Most protozoa are more susceptible to arsenic compounds than are other microorganisms. There are some 24 arsenic compounds listed in the Merck Index [9] which have been or are used in human and veterinary therapeutics. Figure 24 shows the structures proposed for some of these, though few have been crystallographically characterised, Fig. 11. The initial successful use of an arsenic compound, atoxyl, Fig. 24, against sleeping sickness was achieved experimentally by Thomas [3, 6, 197, 215] and introduced by Koch in 1906 [215]. However Paul Ehrlich, the "father of chemotherapy", was the first to realise that the action of the compound was a consequence of its structure [215]. He and his coworkers in the early part of the century synthesised many hundreds of arsenicals, discovering the antisyphilitic agent, Salvarsan or 606 (it was the 606th compound prepared) which was assumed to have the structure shown in Fig. 24, and their work provided the impetus [4] for the synthesis of many more organoarsenic compounds. His contribution to the field has been enumerated by a number of authors including Bauer [215], Albert [6], Schwyzer [216], and Holmstedt et al. [217]. Later developments in the area describing the results of antisyphilitic and anti-nagana testing of several compounds synthesised by the German company Höchst have been outlined [7]. Nowadays these drugs have generally been superseded in the clinic by antibiotics, although a derivative of one of the earlier compounds, melarsoprol, is still used in advanced cases of sleeping sickness, Fig. 24.

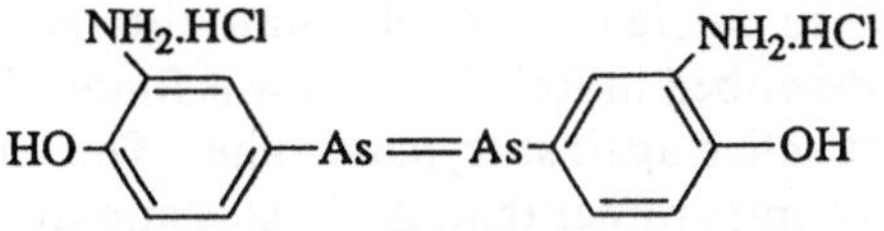

a Arsphenamine ("Salvarsan")
Antisyphilitic

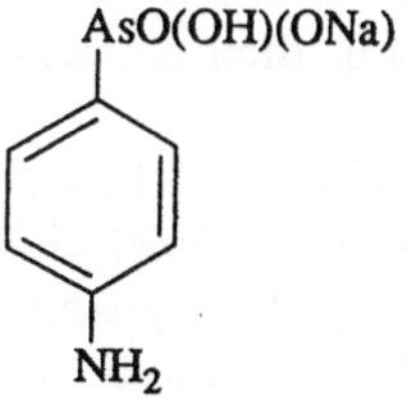

b Atoxyl

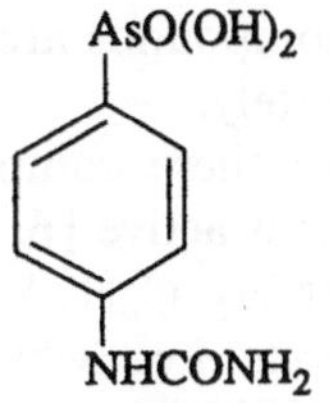

c Carbarsone

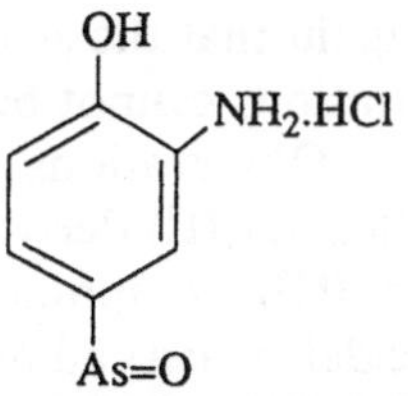

d Oxphenarsine
hydrochloride

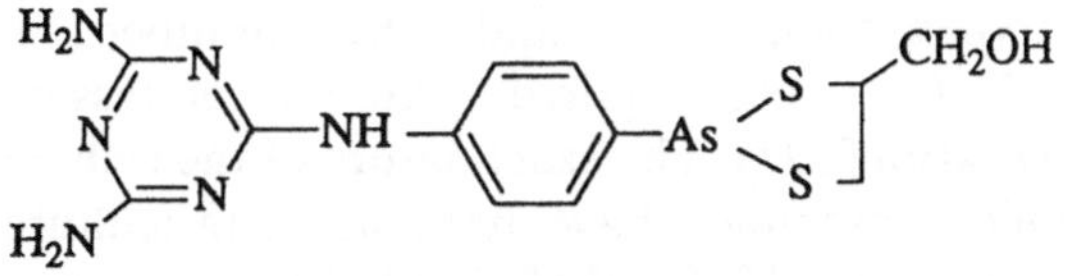

e Melarsoprol
Antitrypanosomal

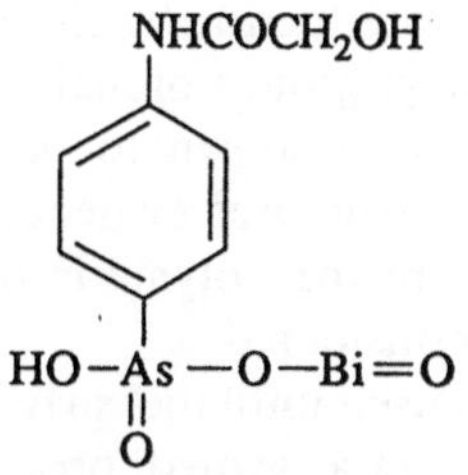

f Glycobiarsol
Antiamoebic

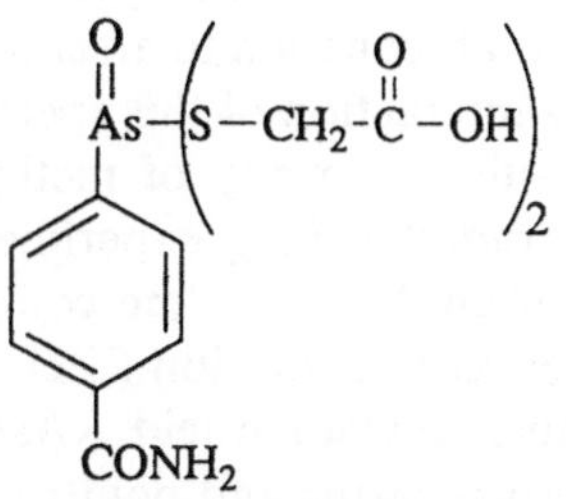

g Arsenamide
Antihelmintic

Fig. 24a–g. Structures proposed for some arsenic compounds which have been used in therapy [9]. Melarsoprol is still used to treat advanced sleeping sickness

Much work has been carried out to elucidate the mechanism of action of these compounds, as described in Ref. [197]. It was found that aromatic arsenic compounds were most active against trypanosomes. The selectivity of arsenical drugs towards trypanosomes rather than the host is apparently conferred by the aromatic ring structure [6] and the substitution on the ring also affects this selectivity [197], e.g. the amide substituted compound $4\text{-}CONH_2\text{-}C_6H_4AsO$ has a parasiticidal/host toxicity ratio of 4.7, compared to 1.22 for $4\text{-}NO_2\text{-}C_6H_4AsO$ [197]. These differences may be due to increased penetration of the amido arseno compound through the protozoal membrane rather than host membranes, however there is no firm evidence for this [197]. It should be noted again that the structures of these compounds are unknown, thus their concentrations cannot be measured accurately.

Observations on the activity of these compounds led to the conclusion that As(III) derivatives were the most active [6, 197]. For example [197], the As(III) compound, $4\text{-}NHCH_2CONH_2\text{-}C_6H_4AsO$, is immediately trypanocidal when administered to rabbits i.v., while the As(V) compound, $4\text{-}NHCH_2CONH_2\text{-}C_6H_4As(O)(OH)(ONa)$ (tryparsamide), has maximum activity only after 6–8 h, suggesting that it must be reduced to As(III) to have an effect. It is usually assumed that the mechanism of action of these compounds involves interaction of As(III) with thiol groups in the trypanosomes [197]. An approximate correlation has been observed for the number of SH groups per trypanosome cell and the number of molecules of $4\text{-}NHCH_2CONH_2\text{-}C_6H_4AsO$ required to kill the cell (assuming a $1:2$ arsenic compound to SH ratio) [197]. The exact nature of the cellular thiols is unclear.

Several arsenic compounds have been used agriculturally as herbicides [218–221] and pesticides [220, 222], Table 14, and many have fungicidal properties [218, 223–225]. Restrictions on the permitted levels of As in the environment have led to a decline in the production and use of these compounds, however in some cases there are no other agents which are as effective [220]. Both As(V) and As(III) compounds have been used and little appears to be known about their detailed mechanisms of action. It has been proposed that the As(V) herbicides, dimethylarsinic acid and its salts, act on Johnsongrass [226] via initial reduction to arsenosomethane (photoreduction by chloroplasts) followed by interaction of this species with sulphydryl groups of malic enzyme. While the site of toxicity of methylarsonate has been shown to be a malic enzyme by radiolabelling experiments there is no structural evidence for the formation of an As–S enzyme complex. Furthermore the polymeric nature of arsenosomethane in solution (Sect. 2.5.1) was not considered.

Derivatives of arsonic acid, $RAs(O)(OH)_2$, were used until the early 1980s as feed additives in swine and poultry, Table 14. They act as growth promoters as well as in the control of enteric diseases [221, 227, 228]. The growth stimulating effect of these arsenic compounds supports the essentiality of arsenic. The beneficial effect of roxarsone on growth was discovered by chance by workers investigating the possible use of this compound against cecal coccidiosis [227]. Other arsenic compounds have been used [9] as spirochetocides in poultry and

Table 14. Uses of arsenic compounds in agriculture and veterinary medicine

Compound	Ox state	Use	Ref.
$Me_2AsSAsMe_2$	III	Pesticide	[222]
4-N-(dimethylaminophenyl)-arsonous dichloride	III	Herbicide	[218]
$NaMeHAsO_3$, Na_2MeAsO_3 mono and di-sodium methylarsonate,	V	Herbicides for grassy and broadleaf weeds	[219]
$Me_2AsO(OH)$ dimethylarsinic acid	V	Herbicide for grassy and broadleaf weeds, desiccant and defoliant in cotton	[219]
H_3AsO_4	V	Desiccant and defoliant in cotton. Wood preservative	[219, 221]
$PbAsO_4$	V	Pesticide used on Florida Grapefruit	[220]
$CaAsO_4$	V	Pesticide to control flies in poultry houses, herbicide to control annual bluegrass on golf courses (U.S.)	[220]
Sodium arsenite	III	Pesticide in sheep and cattle dip	[220]
$(PhAsO)_n$	III	Fungicide	[218]
$(RAsX)_n$ R = alkyl group X = O; n = 1 X = OH, SH; n = 2	III	Fungicide	[223]
2 or 3-NO_2-$C_6H_4As(O)(OH)_2$	V	Fungicide	[218]
10-Phenoxarsine chloride	III	Fungicide Bacteriocide	[224]
10,10'-Oxybis(phenoxarsine)	III	Fungicide e.g. on vinyl films – shower curtains, wall coverings. Bacteriocide	[225]
3-NO_2-4-OH-$C_6H_3As(O)(OH)_2$ "Roxarsone"	V	Growth promoter in poultry, swine. Control of swine dysentry	[221, 227, 228]
4-NH_2-$C_6H_4As(O)(OH)_2$ Arsanilic acid	V	Growth promoter in poultry, swine. Control of swine dysentry	[221, 227, 228]
4-$NH(CONH_2)$-$C_6H_4As(O)(OH)_2$ 4-ureidophenylarsonic acid, "carbarsone"	V	To help in the prevention of blackhead in turkeys	[227]
$NH(COCH_3)$-4-OH-$C_6H_3As(O)(OH)_2$ "Acetarsone"	V	Antihistomonad in turkeys Spirochetocide in poultry	[9]
4-$(CONH_2)$-$C_6H_4As(SCH_2COOH)_2$ "Arsenamide"	III	To treat heartworm infection in dogs	[9]

antihelmintics in dogs. Again little is known about the detailed mechanism by which these compounds act. The choline analogue, arsenocholine, has been shown to have a growth promoting effect in chicks [12]. It is of interest to note that arsenic analogues of choline, acetylarseno- and arsenocholine, have been shown to act as substrates for the enzymes, acetylcholinesterase and choline acetyltransferase [229] from in vitro studies on rat medulla-pons and cerebral cortex and guinea pig myenteric plexus.

3.3.2 Bacterial Resistance

A number of studies on the resistance of bacteria to arsenic have been reported. Many strains of bacteria contain plasmids (extrachromosomal DNA) which confer resistance to a variety of organic and inorganic substances [230, 231]. In *S. aureus* cells, arsenate resistance is due to lowered uptake of arsenate by the inorganic phosphate transport system [231]. Phosphate protects cells from arsenate toxicity, though not from arsenite [230]. It has been shown [232] that pre-incubation of phosphate-starved cyanobacterial cells (*Anabaena variablis*) with arsenate inhibits phosphate transport via a system induced in phosphate-starved cells suggesting that intracellular arsenate is an inhibitor of phosphate in this system. Not surprisingly, the genes coding for arsenate and arsenite resistances are distinct. That for Sb(III) is also probably different [230], though it was found on the same plasmid fragment as that for As(III) [231]. In both *E. coli* and *S. aureus*, resistances to arsenate, arsenite and antimony(III) are encoded by an inducible operon [233] and each ion induces all three resistances. Interestingly, Bi(III) induces arsenate resistance in *E. coli*, though not Bi resistance, while in *S. aureus* there is a genetically separate Bi(III) resistance of unknown mechanism [233]. Bi(III) is a known inducer of the synthesis of the Cys-rich protein metallothionein in eukaryotic cells. Metallothionein could react with As(III), however there appear to be no studies of this possibility.

The reduced accumulation of arsenate in resistant cells is due to an efficient efflux system which allows arsenate excretion via a plasmid-encoded system [234] which is pH-sensitive [230] and ATPase-like [230, 234]. In *E. coli* the arsenate resistance operon for this plasmid system has been cloned [235]. The transport system is in effect an anion-transporting ATPase pump [236], one of the few such known. An arsenite-{as sodium arsenite $(NaAsO_2)_n$} and anti-monite-{as potassium antimony tartrate $K_2[Sb_2(C_4H_2O_6)_2]$} stimulated ATPase-subunit has also been identified [236] and has been shown to consist of two proteins, ArsA and ArsB, coded for by the genes arsA and arsB [236] as well as a third smaller soluble protein which determines substrate specificity [234]. Results reported to date are in agreement with a model involving the ArsA protein as a catalytic energy transducing subunit of a pump [237, 238] while the role of the ArsB protein is as an anion transporting sub-unit [236, 238] and as a membrane binding site for the ArsA protein [238]. There is a structural analogy between this ATPase and an efflux ATPase, P-glycoprotein, linked in mammalian cells with multidrug resistance (i.e. where exposure to one chemotherapeutic agent may cause resistance to other unrelated drugs) and this may indicate possible similar biochemical mechanisms for both these systems [234]. There is some confusion in the literature [239] on the nature of the antimony compound used in the studies outlined in various reports [236, 237, 238], still often described as potassium antimonyl tartrate hemihydrate. This compound is known to be a dimeric Sb(III) species, $K_2[Sb_2(C_4H_2O_6)_2]$, as determined by X-ray crystallography [29] with no Sb=O bonds, and as an Sb(III) species should, by convention, be referred to as antimonite rather than antimonate. The dimeric structure of this compound is retained in aqueous solution as shown by ^{13}C NMR [239], though the influence of pH on the solution structure of the complex has not been studied.

The arsenic(III) cysteine and glutathione compounds, $Me_2As(S-Cys-H)$ and $Me_2As(SG)$, have been found to have bacteriostatic activity in vitro [240]. They totally inhibited growth of the Gram positive bacteria, S. aureus 209, for 15 and 20 mm respectively around the area of sample spotting and the SG complex also showed activity against the Gram negative, K. pneumoniae [240]. No details of doses used were given and since the compounds are unstable in the presence of oxygen, some of the effects may have been due to decomposition products such as cacodylic acid.

Effects of As on Other Cells

In human red blood cells, it has been shown that the sodium pump and the anion exchange transport system accept arsenate as a phosphate analogue, with protein-arsenate interactions similar to those of phosphate [241]. It has also been found that the release of ^{86}Rb from an occluded state of the sodium, potassium pump depends on the relative rates of dephosphorylation and dearsenylation [242].

Exposure of mouse lymphoma cells to elevated temperature (45 °C) or sodium arsenite has been found to induce four nuclear stress proteins (80 to 84 kDa) [243].

3.3.3 Cytotoxicity and Anticancer Activity

Arsenic compounds have been regarded for many years as carcinogens [244–247] however animal models for this are lacking [5, 246, 247] and evidence comes mainly from epidemiological studies [244, 245, 246, 247]. Only in the case of skin cancer does there appear to be a proven link between As exposure and occurrence of the disease [2, 8, 247]. For example, industrial exposure to arsenic (compounds not specified) has led to keratoses of the skin which may become malignant [2]. Also many patients treated with Fowler's solution (daily dose ca. 0.1 mL, i.e. at least 100 mg As_2O_3 [197]) have been reported as having later developed skin cancer [2, 197], though there is some dispute over whether all these cases had in fact been exposed to arsenicals [10]. In a follow-up study (1978) of one of the more famous episodes of arsenic poisoning (in 1943 at St Andrews University, Scotland) where sausage meat was apparently accidentally contaminated with As_2O_3 (from 250 mg to 5 g As/lb), survivors showed no chronic problems attributable to the As exposure [248]. The confusing animal results may arise because As compounds are co-carcinogens rather than primary causes of cancer [5, 173, 247]. The question of the mutagenicity of As compounds also remains unresolved. It is thought [249] that both inorganic As(III) and As(V) are inactive or very weak inducers of gene mutations in vitro and that they are clastogenic and induce sister chromatid exchanges in cells in vitro including human cells; As(III) is more potent in this respect than As(V).

A number of arsenic compounds have been tested for cytotoxic and anti-tumour properties. Among these are the alkylating and structurally related aromatic compounds studied by Bardos et al. [162], Table 15, and As(III) compounds containing the group, SAsMe$_2$ or SeAsMe$_2$ [240, 250], Tables 16, 17 and Fig. 25. The results on the nitrogen-mustard arsenic compounds showed no correlation between alkylating ability and antitumour activity perhaps because the MTD (maximum tolerated dose) was too low. No direct correlation between As oxidation state and activity is apparent either. It has been pro-posed [162] that the observation that the compound RC$_6$H$_4$AsO [R=N(CH$_2$CH$_2$Cl)$_2$] is the most active is best explained by a mechanism of action involving a synergism between the alkylating ability of the R group and the SH-inhibition properties of As(III), i.e. the compound acts at two sites simultaneously. The arseno analogue also was postulated to follow the same mechanism, with an additional in vivo oxidation step. I.r. data indicate [162] that the actual structure of the arsenoso compound involves di- tri- or tetra-meric structures linked via As–O–As bridges rather than containing a formal As=O bond. The uncertainty in structure means that the magnitude of the toxicity figures for these and the arseno compounds should be interpreted with caution.

The compounds listed in Tables 16 and 17 [240, 250] have the structural feature of XAsMe$_2$ (X = S, Se) in common. Tests on XPR$_2$ (X = Se, S; R = Me, Et, Ph) sugar derivatives [250], the P moiety in position R in Fig. 25, showed no

Table 15. Toxicity (μmole/kg) and antitumour activity of arsenic nitrogen mustards and related com-pounds[1]. Data from reference [162]

Compound	p-RC$_6$H$_4$AsO(OH)$_2$	p-RC$_6$H$_4$AsO	p-RC$_6$H$_4$As[S,S'-S$_2$(CH$_2$)$_2$]	(p-RC$_6$H$_4$As)$_n$
R = N(CH$_2$CH$_2$Cl)$_2$				
Toxicity[2]	26	18	38	1280
Activity[3]	−	+ + +	−	+ + +
R = NEt$_2$				
Toxicity	26	12	52	200–2000[4]
Activity	−	+	−	
R = N(CH$_2$CH$_2$OH)$_2$				
Toxicity	3450	28	n	n
Activity	−	−	n	n
R = NH$_2$				
Toxicity	1340	22	n	n
Activity	−	−	n	n

[1] Tested on mice having Ehrlich ascites carcinoma tumours. Also tested against Walker 256 but none of the compounds had an effect on the growth of this tumour. T/C always exceeded 0.8
[2] Approx. LD$_{50}$ values
[3] − , T/C > 0.6 (inactive); + , T/C = 0.2–0.6; + + , T/C = 0.01–0.2; + + + , T/C < 0.01. Where T/C is ratio of mean volume of treated to mean volume of control tumours
[4] LD$_{50}$ for this varied

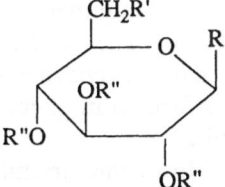

Fig. 25. Structure of the sugar moeity of some AsMe₂ antitumor complexes. Redrawn from Ref. [250]. (R, R', R" defined in Table 17)

Table 16. Antitumour activity of compounds containing the SAsMe₂ or SeAsMe₂ group against mouse tumours. Data from [240]

Compound	Toxicity (mg/kg)	Tumour Type	T/C[1] (dose; mg/kg)	ILS[2] (dose; mg/kg)
Me₂As(S-cys-H)	200	P388	110 (50)	
		L1210		0
Me₂As(S-DL-penicillamine)	50	P388	117 (25)	
		L1210		2 (50)
Me₂As(S-DL-penicillamine) .HCl	100	P388	128 (50)	
		L1210		3 (25)
(Me₂As)(S,S'-homocys-H₂)	> 200	P388	144 (100)	
Me₂As(S-glutathione)	400	P388	164 (100)	
		L1210		33 (500)
Me₂As(S-lipoic acid)	100	P388	128 (50)	
		L1210		33 (100)
Me₂As(S-6-mercaptopurine)	50	P388	144 (25)[3]	
		L1210		7 (25)
Me₂As(Se-cholesterol)	> 200	P388	129 (200)	

[1] T/C: ratio of survival of test to control animals expressed as a percentage
[2] ILS = increased life span
[3] Not reproducible [240]

Table 17. Cytotoxicity and antitumour activity of compounds containing the SAsMe₂ or SeAsMe₂ group[1]. Structures of compounds given in Fig. 25. Data from [250]

Compound				
R	R'	R"	Cell-line (in vitro) KB5 or KB9[2]	Tumour Type (in vivo) 3LE21[3] Walker[4]
OH	SAsMe₂	H	+	+ n
OH	SeAsMe₂	H	±	− n
SAsMe₂	OH	H	+	− n
SeAsMe₂	AcO	AcO	+	− n
Other compounds				
2-(Me₂AsS)-1,3-N₂-C₄H₄			+	− n
Me₂As(S)SAsMe₂			+	− n
Me₂As(O)(OH)			n	− ±

[1] Testing carried out on mice; +, activity confirmed; −, inactive; ±, active but not confirmed; n., not tested in this system
[2] Nasopharyngeal epidermoid carcinoma
[3] Leukemia; mice
[4] Carcinosarcoma; mice

activity in either KB5/KB9 or 3LE21 systems emphasising the importance of the As. The action of the arsenic compounds has been attributed [240] to reaction of enzyme-SH groups with AsR_2 groups causing enzyme deactivation via formation of enzyme-S-AsR_2 bonds. However there is no firm evidence for this.

Inorganic arsenic species such as As_2O_3, and colloidal arsenic have been used in the treatment of some types of cancer [251], though they were not of clinical value. The preparation, Fowler's solution, containing potassium arsenite, was used as an anti-leukaemic agent [197] and it has been noted that in leukaemic blood As is concentrated in the white cells with less in the plasma or red cells [4]. Oxphenarsine hydrochloride has been used as an aid to regression of cancer [10]. It has been observed that some arsenic compounds are specifically antimitotic [197], though the exact mechanism of their carcinostatic properties remains unclear.

Data on the cell toxicity of some metal arsine compounds and ligands [97, 252–254] together with those of P containing analogues are presented in Table 18 and Figs. 26–29. Phosphines and their metal complexes form a novel class of antitumour agents [255]. The metal may act as a carrier for the cytotoxic phosphine ligand protecting it from oxidation until it reaches the target site, or may itself play a role in the activity. The presence of the As in the ligand does not increase toxicity of the ligands or Au(I) complexes in vivo, and there is a parallel reduction in antitumour activity, Table 18 [252, 253] for the system studied, e.g. the ligand dadpe ($Ph_2AsCH_2CH_2PPh_2$) is less toxic than dppe ($Ph_2PCH_2CH_2PPh_2$) by a factor of 1.8 and its antitumour activity is negligible. In vitro studies of Au(I) complexes on other tumour cell lines and

Table 18. Antitumour activity of Group V ligands and metal complexes

Compound	Tumour Type	MTD^1(μmol/kg)	% ILS_{max}^1	Ref
Ligands[2]				
dppe	P388	50	107 ± 4	[252]
dadpe	P388	89	neg	[252]
dpae	P388	82	neg	[252]
Complexes				
[Et$_3$PAuCl]	P388	14	36	[253]
[Et$_3$AsAuCl]	P388	40	18	[253]
[(AuCl)$_2$dppe]	P388	7	98 ± 4	[252]
[(AuCl)$_2$dadpe]	P388	9	neg	[252]
[(AuCl)$_2$dpae]	P388	17	neg	[252]
[Au(dppe)$_2$]Cl	P388	2.9	87	[279]
[Au(dppe)$_2$]Cl	L1210	2.9	≥ 30	[279]

[1] Maximally tolerated dose per day when administered i.p. to B6D2F$_1$ mice on a daily × 5 treatment schedule beginning 24 h after tumour implantation. Maximum increase in life span (ILS equivalent to T/C × 100) produced in a dose response of 4 or 5 dose levels of drug. ILS is the prolongation of median lifespan over that seen in control animals. Compounds with > 30% ILS are usually considered to be active
[2] Abbrev. as dppe, dadpe, dpae for $Ph_2PCH_2CH_2PPh_2$, $Ph_2PCH_2CH_2AsPh_2$, and $Ph_2AsCH_2CH_2AsPh_2$ respectively

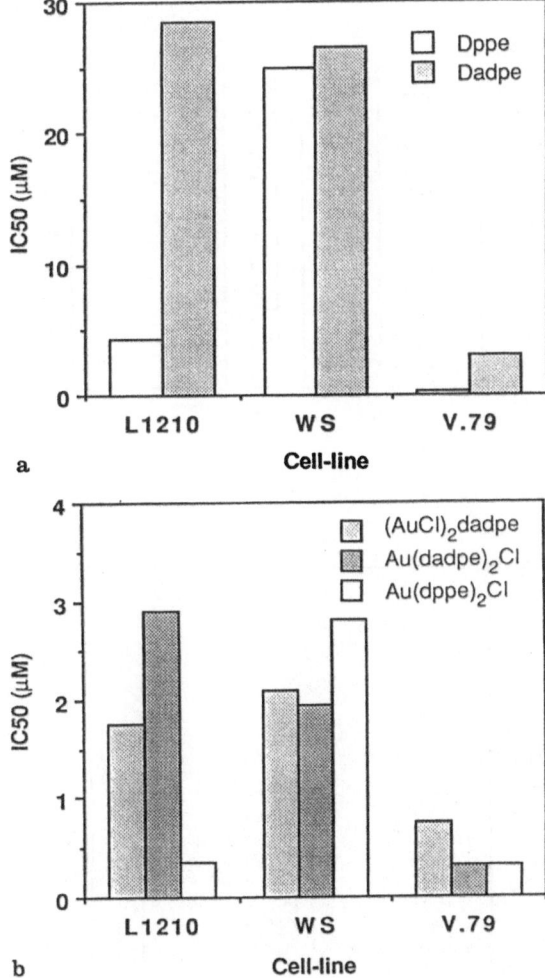

Fig. 26a. Cytotoxicity of dppe and dadpe towards the cell-lines, L1210, WS and V.79 cells where L1210 is antimetabolite sensitive leukaemia, WS are Walker alkylating agent sensitive, and V.79 are Chinese Hamster Lung cells. (**b**) Cytotoxicity of Au(I) complexes of dppe and dadpe towards various cell-lines. Data from [97]

normal V.79 Chinese Hamster Lung cells [97], Fig. 26b, show a significant increase in activity of both dadpe and dppe on complexation to Au(I) e.g. for L1210, the complex [(AuCl)$_2$dadpe] is approx. 16 times more active than the ligand. Furthermore the activity of [Au(dadpe)$_2$]Cl, although less than that of the phosphine analogue, [Au(dppe)$_2$]Cl, in the case of L1210, is significant. Differences in the activity of [Au(dadpe)$_2$]Cl and [Au(dppe)$_2$]Cl may be related to the greater lability of the former in solution [97]. Data for the compounds, [M(L-L)Cl$_2$], [M(L-L)$_2$]Cl$_2$, and [M(L-L)(H$_2$-dmsa)] (where M is Pd, Pt; L-L is dadpe, dppe; H$_2$dmsa is dimercaptosuccinic acid) [254] are shown in Figs. 27–29. It is clear that there is no general increase in cytotoxicity on complexation of the ligands to Pt or Pd and there is no overall trend in activity on changing the ligand from dadpe to dppe, e.g. towards the V.79 cell-line the [Pd(dadpe)Cl$_2$] complex is less active (by a factor of 4.5) than the

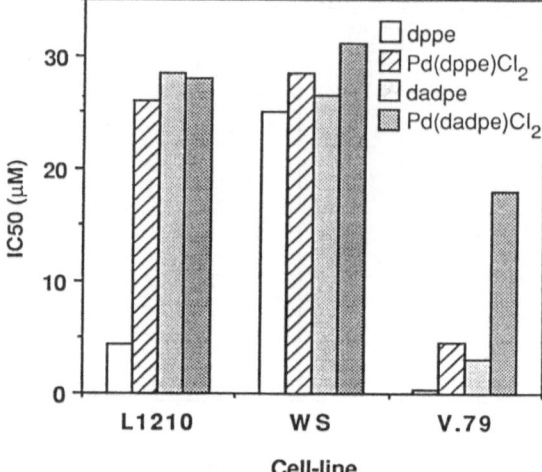

Fig. 27. Cytotoxicity of dadpe and dppe complexes of Pd(II) towards various cell-lines [254]

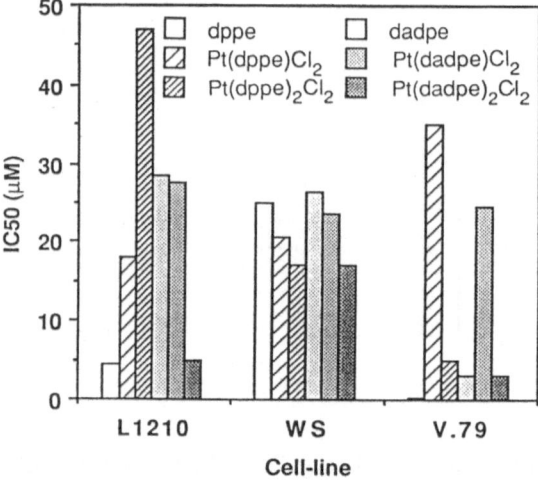

Fig. 28. Cytotoxicity of dadpe and dppe complexes of Pt(II) towards various cell-lines [254]

phosphine analogue, but for the L1210 cell line $[Pt(dadpe)_2]Cl_2$ is *ca.* 10 times more active than $[Pt(dppe)_2]Cl_2$ suggesting that factors such as the nature of the cell-line studied are also important. The other ligands present on the metal are also important as shown by the lack of activity of the dmsa complexes. The mode of action of $[Au(dppe)_2]Cl$ is thought to involve DNA-protein cross-linking and uncoupling of mitochondrial oxidative phosphorylation [256]. Comments on possible mechanisms of action of the arsenic compounds would be speculative at this stage.

A number of Sb and Bi complexes have also been screened for cytotoxicity and antitumour activity. Antimony(III) complexes of ethylenediamine tetra-

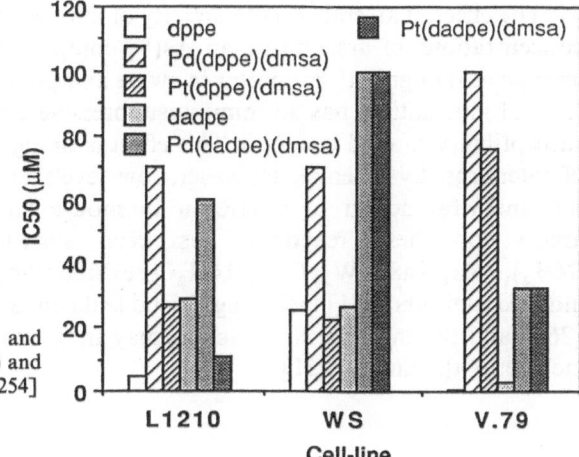

Fig. 29. Cytotoxicity of dadpe and dppe complexes of thiolate Pd(II) and Pt(II) towards various cell-lines [254]

acetic acid, propylenediamine tetraacetic acid, and nitrilotriacetate, cause an increase in lifespan in mice with Ehrlich ascites tumour and spindle cell sarcoma [251]; diphenylantimony(III) thiophosphinates and thiophosphates also have activity against Ehrlich ascites tumour. In vitro, Sb(V) polyamines inhibit HeLa, BHK-21 and L929 cell-lines [251]. A number of Bi(III) compounds have shown in vivo activity, e.g. $Na_2[Bi(O)(6\text{-}mercaptopurine)_3]$ and $[Bi(thioguanine)_3(H_2O)]$ against S180 (Sarcoma 180) and Ca755 (adenocarcinoma 755) in mice [257] and against Dunning ascitic leukaemia in rats [251]. The thiolate complexes, $MeBi(SR)_2$ (R = SMe, 3-thioaniline, $1\text{-}[NH_2Me]^+\text{-}3\text{-}SC_6H_4$) have been reported to have an optimum cure rate of 100% and a therapeutic index of 3.2–5.0 in mice with fluid Ehrlich ascites tumours [258].

3.3.4 Antiviral Activity

Low concentrations of both arsenite and oxphenarsine have been found to inhibit the multiplication of *E. coli* phage without killing the bacteria [197]. An in vitro study of foot-and-mouth disease virus in pig kidney cell culture has shown complete inhibition of the virus at 0.1 mM arsenite [197]. The compound, oxphenarsine hydrochloride, Fig. 24, formerly used as an anti-syphilitic drug [6], has recently been found to have anti-HIV activity [259]. Seven other arsenic compounds were tested including one which was reported to have the same structure as oxphenarsine, but with As(V), but all these were inactive. It was concluded that As(III) is essential for activity. Again it should be stressed that the actual structure of oxphenarsine is not known, and the As(V) compound tested is probably not a structural analogue of this.

The effect of arsenite on the activity of interferon has been studied [5]. High concentrations of arsenite or an As(V) compound, 3-nitro-4-hydroxyphenyl-arsonic acid (a growth promoter in swine and poultry), in drinking water, prior to viral inoculation, has an immunosuppressive effect on mice increasing their susceptibility to viral infection. This effect is thought to be due to an inhibition of interferon by arsenite. However, low levels of arsenic appear to enhance the anti-viral action of interferon in mouse embryo cells [5]. No mechanism for these reactions has been shown. An Sb polytungstate, $(NH_4)_{17}Na[NaSb_9W_{21}O_{86}].14H_2O$, exhibits activity against various virus induced tumours [251] including Friend leukaemia and Moloney sarcoma virus [260] and the mechanism of action may involve inhibition of the viral polymerase by the anion [251].

3.4 Algae and Marine Organisms

A review has recently been published on As speciation in the envionment [12] and from this the ubiquity of As in all its forms is clear. Organoarsenic compounds have been isolated from a variety of marine animals including crustacea such as the Western rock lobster (*Panulirus longipes cygnus* George – the first species from which a characterised As compound was isolated [53, 261]), various clams, and shrimps [12], as well as fish, e.g. North Sea plaice, school whiting, haddock, mackerel and salmon [12]. In addition, arsenic compounds have been found in marine algae, e.g. Brown Kelp [262] and in some edible seaweeds in Japan [263]. It is interesting to note that in one of these seaweeds, Hiziki (*Hizikia fusiforme*), the major form of arsenic is an inorganic arsenic species [71.8 p.p.m. of dry matter; As(III) and As(V)], yet no cases of poisoning or death due to Hiziki appear to have been reported [263]. However, later data [12] indicate that the arsenic is present as arseno-sugars and arsenate; the previously isolated arsenite may have been a degradation product of these arising because of the harsh extraction conditions used [12]. The highest concentration of As in any species is between 6000 and 13000 p.p.m. (i.e. average concentrations of 0.08–0.17 M!) found in the polychaete worm, *Tharyx marioni* [12]. Thus there is a variation within the same species in total As concentration. The concentrations observed for marine animals where the compounds have been isolated range from 0.31 p.p.m. in salmon muscle to 340 p.p.m. in the mid gut gland of the gastropod *Charonia sauliae* [12].

Natural levels of dissolved As in sea-water [264] are between 1000 and 2000 ng/L (13–26 nM) usually predominantly as arsenate with some arsenite (as $As(OH)_3$) and methylated acids, $MeAs(O)(OH)_2$, $Me_2As(O)OH$, these three species arising from biological activity. Similar levels are present in unpolluted rivers; however the average concentration in European rivers in industrial and urban areas is 3500 ng/L (47 nM) [264].

Table 19. Naturally occurring organoarsenic compounds isolated from marine organisms

Compound	Occurrence	Probable Origin
Arsenobetaine $(CH_3)_3As^+CH_2COO^-$	Most marine animals [12] First isolated from Western Rock Lobster [261]	See text
Me_3AsO	Plaice, estuary catfish [188]	Bacterial action on ingested As compounds in the gut
Me_3As	Prawns [188]	Bacterial action on ingested As compounds in the gut
$[Me_4As]X$	Gill of the Clam [146]	Possibly from plankton
2-R-3-R'-5-deoxy-5-(dimethylarsinoyl)-β-ribofuranoside[1]	Brown Kelp [268]	Intermediates in the biosynthesis of arsenobetaine
2-R-3-R'-5-deoxy-5-(dimethylarsinoyl)-β-ribofuranoside[2]	Kidney of Giant Clam [268]	Metabolic waste products of unicellular algae present in clam tissue
Arsenocholine $[(CH_3)_3AsCH_2CH_2OH]X$	Shrimp [271][3]	
Dimethyloxarsylethanol $(CH_3)_2As(O)CH_2CH_2OH$	Brown Kelp [270]	Anerobic decomposition of brown kelp

[1] Three compounds isolated with R = hydroxy, R' = sulphopropyl; R = R' = dihydroxypropyl; R = hydroxy, R' = $[PO_3CH_2CH(OH)CH_2OH]^-$
[2] Two compounds isolated with R = R' = dihydroxypropyl; R = hydroxypropyl, R' = OSO_3H
[3] Other workers failed to find this [12]

Table 19 shows the arsenic compounds which have been isolated from algae and marine animals; the usual method of characterisation involves comparison of properties with those of synthetic analogues. The organoarsenic compound, arsenobetaine, is found in almost all marine animals consumed by man [265] and is generally the most abundant, a notable exception being the sea squirt (*Halocynthia roretzi*) [12]. Thus, there is much interest in the origin of arseno-betaine [12, 265, 266, 267]. Experimental evidence to date suggests that arsenate in the seawater is initially taken up and metabolised by algae [266, 265] to dimethyl(ribosyl)arsine oxides, and these metabolites are deposited in the sediments, i.e. benthos, where microbial action converts them to arsenocholine and arsenobetaine precursors. From the sediments these are transferred by an unknown mechanism to fish and crustaceans. A possible mechanism [266] is shown in Fig. 30. Proposed intermediates are indicated as such. The arsenic containing ribofuranosides were first isolated [262, 268] from the algae, *Ecklonia radiata*, and have since been found in the Giant Clam, *Tridacna maxima* (as metabolites of unicellular zooxanthellae present in the clam gut [54]), and in *Laminaria japonica* and *Hizika fusiforme* [269]. Thus these are probably common intermediates of algal metabolism of arsenate. The proposed

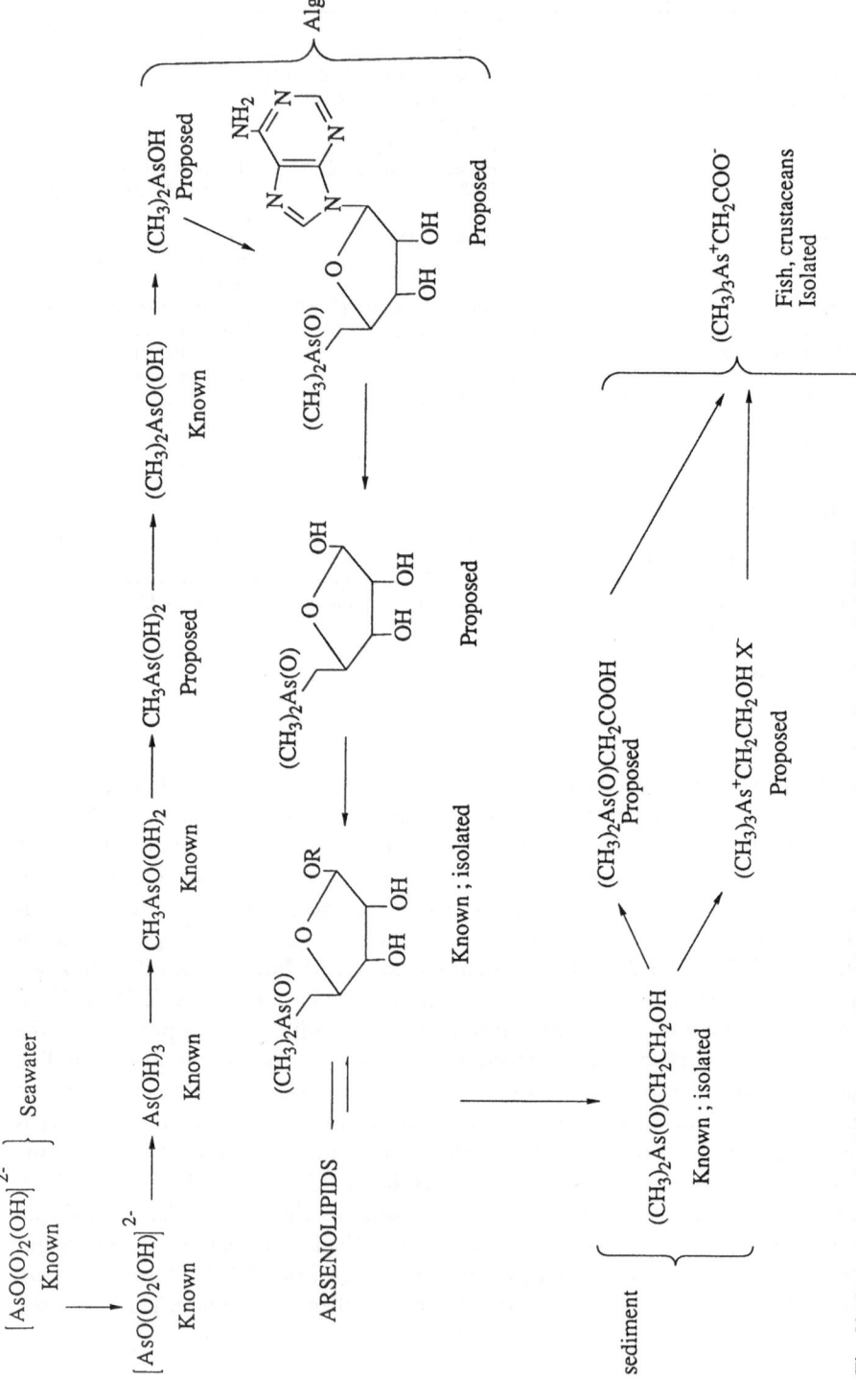

Fig. 30. Proposed mechanism for conversion of arsenate to arsenobetaine. Based on Ref. [266]

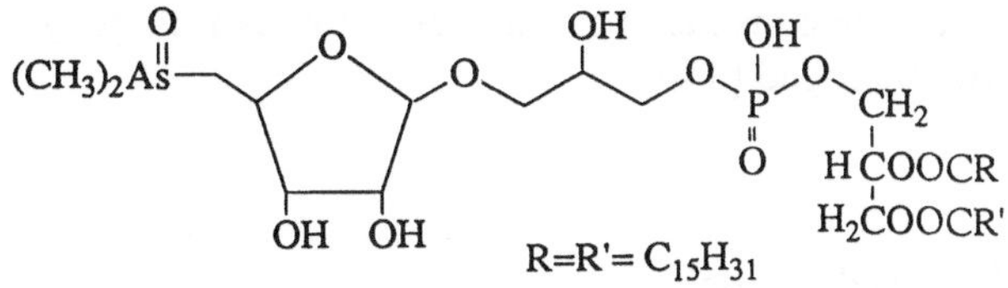

Fig. 31. Proposed structure of a lipid soluble arsenic compound. Redrawn from Ref. [12].

$(CH_3)_2\overset{\overset{O}{\|}}{As}$ — ... OH OH ... $R=R'=C_{15}H_{31}$

mechanism of formation of these is shown in Fig. 30. However the key inter-mediate in which the adenosyl is bound to the ribose ring has not been observed to date.

Lipid-soluble arsenic compounds have been isolated from a number of organisms [12, 266, 268], however only one such structure has been tentatively characterised, Fig. 31, from Wakame or *Undaria pinnatifida* [12]. The nature of this structure supports the role of arsenosugars in Fig. 30. Anaerobic decomposition of two of these sugars has been shown to give dimethyloxarsylethanol, $(CH_3)_2As(O)CH_2CH_2OH$ [270]. The occurrence of arsenocholine in shrimps has been reported by some workers [271], whereas others have not found it [12]. The mechanism of the final stages in the formation of arsenobetaine involving the addition of a third CH_3 group to arsenobetaine is unknown [266]. Methylation pathways in microorganisms and fungi do not lead to the formation of quaternary arsonium salts, Sect. 3.1.2. Degradation of arsenobetaine by sedimentary microorganisms to Me_3AsO and an unidentified arsenic compound, possibly monomethylated, has been shown [272].

Freshwater organisms accumulate arsenic to a lesser extent than marine organisms allowing for differences in sea and fresh water arsenic concentrations [273] and no arsenobetaine has been found in them [273]. The proposal that only algae metabolise arsenate and higher animals do not, is not general for all species [12]. The snails, *Littorina littoralis*, and *Nucella lapillus*, appear to synthesise a water soluble and a lipid soluble arsenic compound from arsenate [12]. The toxicology of arsenobetaine has been investigated in rats [274], rabbits [274], and mice [274, 275] and it is excreted unchanged after a couple of days, mainly in the urine and no other As containing species are present. Thus arsenobetaine is not biotransformed in these animals. Neither arsenobetaine nor arsenocholine are embryotoxic [276]. It has been shown that arsenocholine is oxidised to arsenobetaine prior to excretion [12] in mammals. Arsenobetaine is excreted unchanged in human urine after ingestion of cooked Western rock lobster [53]. Early experiments [12] on the addition of arsenocholine to chick or rat liver extracts showed the release of a gas, possibly Me_3As, while oral administration of arsenobetaine to these resulted in emission of a garlic-like odour from the tissues and exhalations of both rats and chicks – ascribed to Me_3As. These results suggest that organoarsenic compounds are metabolised in some animals, probably by intestinal flora [12]. Apart from arsenobetaine a number of other As-containing compounds have been isolated, some of which have not been well characterised, thus their toxicity is unknown. In particular the toxicity of the tetramethylarsonium salt found in the clam, *Meretrix lusoria*, should be considered in view of the known toxicity of the tetramethylammonium analogue [146].

4 Conclusions

Arsenic is a potentially essential element for warm-blooded mammals and is abundant in some other creatures, especially certain marine organisms. Most of the latter contain methylated As(V) derivatives especially arsenobetaine, $Me_3As^+CH_2CO_2^-$, and also arsenosugars. The rôle of arsenobetaine is unclear, but so too is the rôle of the parent nitrogen analogue betaine in man. Betaine is an intracellular metabolite in the kidney, thought to protect proteins against denaturation by urea [277]. It is only excreted in the early stages of life (until about 2 years of age) when kidney development is complete. Perhaps arsenobetaine plays a similar rôle in protecting proteins in marine organisms. When consumed in sea-food, arsenobetaine appears to be readily excreted unchanged by man, and therefore is non-toxic. The formation of arsenobetaine is a detoxifying route for inorganic arsenic.

Pathways for the methylation of arsenic (detoxification pathways) also exist in man and microorganisms, involving CH_3^+ donors such as S-adenosylmethionine (SAM). This contrasts with P metabolism, since P does not appear to be methylated in vivo, probably a reflection of the ease of reducing As(V) to As(III) compared to P. Phosphorus(V) reduction seems to occur only in some anaerobic bacteria [255, 278], presumably with the aid of ferredoxins.

Aryl arsonic acids $RAs(V)(O)(OH)_2$ have been used as growth promoters for pigs and poultry, but a systematic search for compounds with optimal activity does not appear to have been made, and the molecular basis for growth promotion is unknown. Perhaps it involves stimulation of metallothionein synthesis and a feed-back into zinc metabolism. High levels of one of these compounds, $3\text{-}OH\text{-}4\text{-}NO_2\text{-}C_6H_3As(V)(O)(OH)_2$, in drinking water have been reported to have an immunosuppressive effect on mice.

Although arsenic compounds were widely used as antimicrobial agents earlier this century, the structures of many of them are still uncertain. The chemistry of arsenic is complicated by the lack of good probes, for example ^{75}As NMR is much less useful than ^{31}P NMR since it is strongly quadrupolar, there is a tendency to form polymeric species involving As–O–As-linkages, and some of the biologically-relevant redox chemistry is not well-understood.

Some other contrasts between arsenic and phosphorus are notable. Arsenious acid exists at biological pH (7) as $As(OH)_3$ rather than $HAs(O)(OH)$ or an arsenite anion, c.f. $[HP(OH)O_2]^-$, and in the solid-state arsenite salts can be polymeric. Arsenate, As(V), esters are much more labile than phosphate esters and therefore less useful for energy storage in cells. It is possible that the formation of nucleotide arsenates is involved in the antiviral (HIV) activity of the As(III) arsenoso compound oxphenarsine. Arsenate incorporation followed by reduction and methylation might lead to termination of the reverse transcription of viral RNA. Further investigation is needed of oxphenarsine, the structure of which is unknown, even though the compound was first synthesised over 70 years ago. It is probably polymeric with bridging oxygens and a cyclic structure.

It is remarkable that despite the widely-quoted ability of As(III) to inhibit thiol- and dithiol-containing enzymes, few detailed structural investigations of As-thiol linkages have been carried out. Indeed, there appear to be no X-ray structures of arsenic-containing proteins or peptides. Sometimes proteins for X-ray work are crystallised from cacodylic buffers, and cacodylate is readily reduced to As(III) by thiols. There may be some cases where As has been incorporated into proteins during crystallisation. Compounds such as the product from reaction of cacodylic acid with the thiol glutathione, Me_2AsSG are reported to exhibit antileukaemic activity, but since they are relatively unstable in air this may have complicated the testing.

Metal–arsine bonds tend to be more labile than metal–phosphine bonds, but this could have some advantage in anticancer agents which use metals to deliver toxic ligands to tumour cells. The apparent ability of cells to metabolise phosphines and arsines differently warrants further investigation. In general there appears to have been little systematic testing of arsenic compounds for anticancer activity.

Acknowledgements. We thank the Cancer Research Campaign, Medical Research Council, Science and Engineering Research Council, Wolfson Foundation and University of London for their support for our work. We are also grateful to Professor K. Irgolic (University of Graz), Dr. B.W. Fitzsimmons (Birkbeck College), Dr. J.K. Chesters (Rowett Research Institute), and Dr. N. Price (Slough) for helpful discussions about various aspects of this subject, and Professor B.J. Aylett (Queen Mary College) for helpful comments on the script.

5 References

1. Polson CJ, Green MA, Lee MR (1983) Clinical Toxicology, 3rd edn, Pitman, London, chap 23, p 421
2. Schroeder HA, Balassa JJ (1966) Journal of Chronic Diseases *19*: 85 and references cited therein
3. Doak GO, Freedman LD in: Medicinal Chemistry (ed A Burger), 2nd edn, Interscience, New York 1960, chap 47, pp 1027–1049 and references cited therein
4. Vallee BL: AMA Archives of Industrial Health *21*, 132 (1960) and references cited therein
5. Squibb KS, Fowler BA (1983) In: Fowler BA(ed) Biological and environmental effects of arsenic, Elsevier, Amsterdam, chap 7, p 233 and references cited therein
6. Albert A: Selective Toxicity, the physicochemical basis of therapy, 7th edn, Chapman and Hall, London 1985, (a) chap 13, pp 551–555 (b) chap 6, pp 206–219
7. Herrmann W, Hilmer H: Angew Chem *66*, 349 (1954)
8. Hindmarsh JT, McCurdy RF: CRC Critical Reviews in Clinical Laboratory Sciences *23*, 315 (1986) and references cited therein

9. Bunclavari S (ed) (1989) The Merck Index 11th edn, Merck, New Jersey
10. Frost DV in: Inorganic and Nutritional Aspects of Cancer (ed Schrauzer GN), Plenum, New York 1978, chap 18, pp 259–279
11. Datta DV, Narang APS: Bulletin of the Postgraduate Institute of Med Educ Research, Chandigarh *13*, 180 (1979)
12. Cullen WR, Reimer KJ: Chem Rev *89*, 713 (1989)
13. Knowles FC, Benson AA: Trends in Biochemical Science, 178 (1983)
14. Greenwood NN, Earnshaw A: Chemistry of the Elements, Pergamon, Oxford 1984, chaps 12 (P) and 13 (As, Sb, Bi)
15. Santhanam KSV, Sundaresan NS: in Standard Potentials in Aqueous Solution (eds Bard AJ, Parsons R, Jordan J), Marcel Dekker, New York 1985, chap 7, part III, pp 162–172
16. Santhanam KSV in: Standard Potentials in Aqueous Solution (eds Bard AJ, Parsons R, Jordan J), Marcel Dekker, New York 1985, chap 7, part II, pp 139–162; Past V in: ibid, part IV, pp 172–179
17. Phillips CSG and Williams RJP: Advanced Inorganic Chemistry, Oxford University Press, Oxford 1965, chap 17, p 632
18. Stanbury DM: Adv Inorg Chem *33*, 69 (1989)
19. Doak GO, Freedman LD: Organometallic compounds of arsenic, antimony, and bismuth, Wiley, New York 1970 (a) chap 2 and references cited therein (b) chap 3 and references cited therein
20. Watson A, Svehla G: Analyst (London) *100*, 489 (1975) and references cited therein
21. Brdicka R: Coll Czech Chem Commun *7*, 457 (1935)
22. Lowry JH, Smart RB, Mancy KH: Anal Chem *50*, 1303 (1978) and references cited therein
23. DuPont TJ, Mills JL: Inorg Chem *12*, 2487 (1973)
24. Huttner G, v Seyerl J, Marsili M, Schmid HG: Angew Chem Int Ed Eng *14*, 434 (1975)
25. Huttner G, Muller HD, Frank A, Lorenz H: Angew Chem Int Ed Eng *14*, 705 (1975)
26. v Seyerl H, Moerins H, Wagner A, Frank A, Huttner G: Angew Chem Int Ed Eng *17*, 844 (1978)
27. Kuykendall GL, Mills JL: J Organomet Chem *118*, 123 (1976)
28. Matsumoto K, Kawaguchi H, Kuroya H, Kawaguchi S: Bull Chem Soc Japan *46*, 2424 (1973)
29. Kamenar B, Grdenic D, Prout CK: Acta Crystallogr *B26*, 181 (1970)
30. Wells AJ: Structural Inorganic Chemistry, 5th edn, Clarendon, Oxford 1984, (a) chap 19, p 844 (b) chap 20, p 866 and references cited therein
31. Raston CR, White AH: J Chem Soc Dalton Trans, 2425 (1975)
32. Raston CR, White AH: J Chem Soc Dalton Trans, 791 (1976)
33. Pertlik P: Monatsheft für Chemie *106*, 755 (1975)
34. Menary JW: Acta Cryst *11*, 742 (1958)
35. Corbridge DEC: Acta Cryst *9*, 991 (1956)
36. Doak GO, Long GG, Freedman LD in: Kirk-Othmer, Encyclopaedia of Chemical Technology (Vol 3), 1982, pp 251–266
37. Jones DEH, Ledingham KWD: Nature *299*, 626 (1982)
38. Aylett BJ: Organometallic compounds, The Main Group Elements (Vol 1, Part Two), Groups IV and V, 4th edn, Chapman and Hall, London 1979, chap 7, pp 390–507
39. Arif AM, Cowley AH, Pakulski M: J Chem Soc Chem Commun, 165 (1987)
40. Rheingold AL, DiMaio AJ: Organometallics *5*, 393 (1986)
41. Camerman N, Trotter J: J Chem Soc, 219 (1964)
42. Cordes AW, Gwinup PD, Malmstrom MC: Inorg Chem *11*, 836 (1972)
43. Dräger M: Z Anorg Allg Chem *411*, 79 (1975)
44. Bergerhoff G, Namgung H: Zeitschrift für Kristallographie *150*, 209 (1979)
45. Cotton FW, Wilkinson G: Advanced Inorganic Chemistry, 4th edn, Wiley-Interscience, Chichester 1980, chap 14
46. Holmes RR: Acc Chem Res *12*, 257 (1979)
47. Holmes RR: Prog Inorg Chem *32*, 119 (1984)
48. Holmes RH, Day RO, Sau AC: Organometallics *4*, 714 (1985) and references cited therein
49. Worzala H: Acta Cryst *B24*, 987 (1968)
50. Mighell AD, Smith JP, Brown WE: Acta Cryst *B25*, 776 (1969)
51. Cullen WR: Adv Organomet Chem *4*, 145 (1966) and references cited therein
52. Kostick A, Secco AS, Billinghurst M, Abrams D, Cantor S: Acta Cryst *C45*, 1306 (1989)
53. Cannon JD, Edmonds JS, Francesconi KA, Raston CL, Saunders JB, Skelton BW, White AH: Aust J Chem *34*, 787 (1981)

54. Edmonds JS, Francesconi KA, Healy PC, White AH: J Chem Soc Perkin Trans I, 2989 (1982)
55. Shimada A: Bull Chem Soc Japan 34, 639 (1961)
56. Chatterjee A, Gupta SPS: Acta Cryst B33, 3593 (1977)
57. Chatterjee A, Gupta SPS: Acta Cryst B33, 164 (1977)
58. Stuckey JE, Cordes AW, Handy LB, Perry RW, Fair CK: Inorg Chem 11, 1846 (1972)
59. Bohra R, Roesky HW, Noltemeyer M, Sheldrick GM: J Chem Soc Dalton Trans, 2011 (1984)
60. Dove MFA, Sowerby DB: Coordination Chem Rev 75, 297 (1986)
61. Sigwarth B, Zsolnai L, Scheidsteger O, Huttner G: J Organomet Chem 235, 43 (1982)
62. Ashe AJ III, Ludwig EG Jr: J Organometallic Chem 303, 197 (1986) and references cited therein
63. Hedberg K, Hughes EW, Waser J: Acta Cryst 14, 369 (1961)
64. Rheingold AL, Sullivan PJ: Organometallics 2, 327 (1983)
65. Smith LR, Mills JL, J Organometal Chem 84, 1 (1975) and references cited therein
66. Levinson AS: J Chem Ed 54, 98 (1977)
67. Burns JH, Waser J: J Amer Chem Soc 79, 859 (1957)
68. Cowley AH, Pinnel RP: Topics in Phosphorus Chemistry 4, 1 (1967)
69. Smith LR, Mills JL: J Amer Chem Soc 98, 3852 (1976)
70. DuPont TJ, Smith LR, Mills JL: J Chem Soc Chem Commun, 1001 (1974)
71. Breunig HJ, Häberle K, Dräger M, Severengiz T: Angew Chem Int Ed Eng 24, 72 (1985)
72. Daly JJ, Sanz, F: Helv Chim Acta 53, 1879 (1970)
73. Dimaio AJ, Rheingold AL: Chem Rev 90, 169 (1990)
74. Cowley AH, Lasch JG, Norma NC, Pakulski M: J Amer Chem Soc 105, 5506 (1983) and references cited therein
75. Cowley AH, Norman NC: Prog Inorg Chem 34, 1 (1986) and references cited therein
76. Huttner G: Pure and Appl Chem 58, 585 (1986) and references cited therein
77. Carty AJ, Taylor NJ, Coleman AW, Lappert MF: J Chem Soc, Chem Commun, 639 (1979)
78. Drew MGB, Wolters AP, Tomkins IB: J Chem Soc Dalton Trans, 974 (1977)
79. Foy RM, Kepert DL, Raston CL, White, AH: J Chem Soc Dalton Trans, 440 (1980)
80. Drew MGB, Wolters AP: Acta Cryst B33, 205 (1977)
81. Gilli G, Sacerdoti M, Domiano P: Acta Cryst B30, 1485 (1974) and references cited therein
82. Cullen WR, Dolphin D, Einstein FWB, Mihichuk LM, Willis AC: J Amer Chem Soc. 101, 6898 (1979)
83. Busetto C, D'Alfonso A, Maspero F, Perego G, Zazzetta A: J Chem Soc Dalton Trans, 1828 (1977)
84. Mague JT: Inorg Chem 12, 2649 (1973)
85. Mague JT: Inorg Chem 9, 1610 (1970)
86. Manojlovic-Muir L, Muir KW, Ibers JA: Discussions of the Faraday Society 47, 84 (1969)
87. Dunn JBR, Jacobs R, Fritchie CJ Jr: J Chem Soc Dalton Trans, 2007 (1972)
88. Spek AL, van Eijch BP, Jans RJF, van Koten G: Acta Cryst C43, 1878 (1987)
89. Pierpoint CG, Eisenberg, R: Inorg Chem 11, 828 (1972)
90. Holloway RG, Penfold BR, Colton R, McCormick MJ: J Chem Soc, Chem Commun, 485 (1976)
91. Colton R, McCormick MJ, Pannan CD: Aust J Chem 31, 1425 (1978)
92. Cheng PT, Nyburg SC: Can J Chem 50, 912 (1972)
93. Russell DR, Tucker PA: J Chem Soc Dalton Trans, 1752 (1975)
94. Gill JT, Mayerle JJ, Welcker PS, Lewis DF, Ucko DA, Barton DJ, Stowens D, Lippard SJ: Inorg Chem 15, 1155 (1976)
95. Puddephatt RJ: The Chemistry of Gold, Elsevier, Amsterdam 1978, pp 261–269 and references cited therein.
96. Berners-Price SJ, Mazid MA, Sadler PJ: J Chem Soc Dalton Trans, 969 (1984)
97. Ni Dhubhghaill OM, Sadler PJ, Kuroda R: J Chem Soc Dalton Trans, 2913 (1990)
98. Critical Stability Constants (eds Smith RM, Martell AE), Plenum, vol 4, pp 55–56 (1976), vol 5, pp 409–410 (1982)
99. Merijanian A, Zingaro RA: Inorg Chem 5, 187 (1966)
100. Cooney RV, Mumma RO, Benson AA: Proc Nat Acad Sci U.S.A 75, 4262 (1978)
101. Data for Biochemical Research (eds Dawson RMC, Elliot DC, Elliot WH, Jones KM), 3rd edn, Clarendon Press, Oxford 1986, p 9
102. Pauling L: The Nature of the Chemical Bond, 3rd edn, Cornell University Press, New York 1960, p 85
103. Ananthakrishnan R: Proceedings of the Indian Academy of Science 5A, 200 (1937)
104. Baudler M: Z Elektrochem 59, 173 (1955); Chem Abs 49, 11418h (1955)

105. Van Wazer JR, Callis CF, Shoolery JN, Jones RC: J Amer Chem Soc 78, 5715 (1956)
106. Furberg S, Landmark P: Acta Chem Scand 11, 1505 (1957)
107. Dos Remedios Pinto MAF, Sadler PJ, Neidle S, Sanderson MR, Subbiah A, Kuroda R: J Chem Soc Chem Comm 13 (1980)
108. Loehr TM, Plane RA: Inorg Chem 7, 1708 (1968)
109. Kirschenbaum LJ, Rush JD: Inorg Chem 22, 3304 (1983)
110. Baer CD, Edwards JO, Rieger PH, Silva CM: Inorg Chem 22, 1402 (1983)
111. Copenhafer WC, Rieger PH: J Amer Chem Soc 100, 3776 (1978)
112. Okumura A, Yamamoto N, Okazaki N: Bull Chem Soc Japan 46, 3633 (1973)
113. Okumura A, Okazaki N: Bull Chem Soc Japan 46, 2937 (1973)
114. Dennard AE, Williams RJP: J Chem Soc A, 812 (1966)
115. Baer CD, Edwards JO, Kaus MJ, Richmond TG, Rieger PH: J Amer Chem Soc 102, 5793 (1980)
116. Baer CD, Edwards JO, Rieger PH: Inorg Chem 20, 905 (1981)
117. Lagunas R: Arch Biochem Biophys 205, 67 (1980)
118. Mayes PA in: Harper's Review of Biochemistry (20th edn eds Martin DW, Mayes PA, Rodwell VW, Granner DR), Lange, Los Altos 1983, p 170
119. Brill TR, Campbell NC: Inorg Chem 12, 1884 (1973)
120. Kolling OW, Mawdsley EA: Inorg Chem 9, 408 (1970)
121. Huheey JE: Inorganic Chemistry, 2nd edn., Harper and Row, New York 1978, Chap 4, 162
122. Gamayurova VS: Uspekhi Khimii 50, 1601 (1981) [Engl trans: Russian Chem Rev 50, 836 (1981)]
123. Laskorin AN, Yakshin VV, Lyubosvetova, NA: Uspekhi Khimii 50, 860 (1981) [Engl trans: Russian Chem Rev 50, 454 (1981)]
124. Ahrland S, Chatt J, Davies NR: Quart Rev Chem Soc 12, 265 (1958)
125. Ahrland S, Berg T, Trinderup P: Acta Chem Scand A31, 775 (1977)
126. Ahrland S, Hultén F, Persson I: Acta Chem Scand A40, 595 (1986)
127. Ahrland S, Berg T, Bläuenstein P: Acta Chem Scand A32, 933 (1978)
128. Hancock RD, Marsicano F: J Chem Soc Dalton Trans., 1832 (1976)
129. Roulet R, Lan NQ, Mason WR, Fenske Jr GP: Helv Chim Acta 56, 2405 (1973)
130. Kidd RG in: Multinuclear Approach to NMR spectroscopy, NATO ASI Series, Series C, vol 103 (eds Lambert JB, Riddell FG), D Reidel, Dordrecht 1983, Chap 18, pp 379–387
131. Harris RK in: NMR and the Periodic Table (eds Harris RK, Mann B), Academic Press, London 1978, Chap 11, pp 379–382 and references cited therein
132. Balimann G, Pregosin PS: J Magn Res 26, 283 (1977)
133. Brevard C, Granger P in: Handbook of High Resolution Multinuclear NMR, Wiley, N.Y. 1981, p 90(N), p 102(P), p 136(As), p 170(Sb), p 210(Bi)
134. Minkwitz R, Prenzel H, Schardey A, Oberhammer H: Inorg Chem 26, 2730 (1987)
135. Irgolic KJ (personal communication)
136. Knoll F, Marsmann HC, Van Wazer JR: J Amer Chem Soc 91, 4986 (1969)
137. Elmes PS, Middleton S, West BO: Aust J Chem 23, 1559 (1970)
138. Marsmann HC, Van Wazer JR: J Amer Chem Soc 92, 3969 (1970)
139. Durand M, Laurent JP: J Organomet Chem 77, 225 (1974)
140. Baudler M, Carlsohn B, Böhm W, Reuschenbach G: Z Naturforsch 31b, 558 (1976)
141. Pouchert CJ: Aldrich Library of NMR spectra, 2nd edn, vol 2, Milwaukee, Aldrich Chemical Co 1983
142. Schiemenz GP: J Organomet Chem 52, 349 (1973)
143. Hendrickson JB, Maddox ML, Sims JJ, Kaesz HD: Tetrahedron 20, 449 (1964)
144. Verstuyft AW, Cary LW, Nelson JH: Inorg Chem 14, 1495 (1975)
145. Randall EW, Shaw D: Spectrochim Acta A23, 1235 (1967)
146. Shiomi K, Kakehashi Y, Yamanaka H, Kikuchi T: Appl Organomet Chem 1, 177 (1987)
147. Massey AG, Randall EW, Shaw D: Spectrochim Acta A21, 263 (1965)
148. Brown JCC: Ph.D Thesis, University of London (1987)
149. Edwards RG, Hands AR: Biochim Biophys Acta 431, 303 (1976)
150. Irgolic KJ, Junk T, Kos C, McShane WS, Pappalardo GC: Appl Organomet Chem 1, 403 (1987)
151. Cushley RJ, Mautner HG: Tetrahedron 26, 2151 (1970)
152. Schmidbaur H, Nusstein P: Organometallics 4, 344 (1985)
153. Buslaev YA, Kravcenko EA, Kolditz L: Coord Chem Rev 82, 7 (1987) and references cited therein.

154. Zakirov DU, Safin IA: Khim Tekhnol Elementoorg Soedin Polim, 36 (1981); Chem Abs 96, 191811c (1981)
155. Bastow TJ, Elmes PS: Aust J Chem 27, 413 (1974)
156. Brill TB, Parris GE, Long GG, Bowen LH: Inorg Chem 12, 1888 (1973)
157. Crow JP, Cullen WR, Herring FG, Sams JR, Tapping RL: Inorg Chem 10, 1616 (1971)
158. Mosbæk, H: Acta Chem Scand A29, 957 (1975)
159. Bowen LH, Garrou PE, Long GG: Inorg Chem 11, 182 (1972)
160. Brown SSD, Salter ID, Dyson DB, Parish RV, Bates PA, Hursthouse MB: J Chem Soc Dalton Trans., 1795 (1988) and references cited therein
161. Nakamoto K: Infrared Spectra of Inorganic and Coordination Compounds, 2nd edn., Wiley-Interscience, New York 1970, Section II, pp 78–148
162. Bardos TJ, Datta-Gupta N, Hebborn P: J Med Chem 9, 221 (1966)
163. Pouchert CJ: Aldrich Library of Infrared spectra, 3rd edn., Milwaukee Aldrich Chemical Co 1981, 1557G
164. Brooks RR, Ryan DR, Zhang H: Anal Chim Acta 131, 1 (1981)
165. Cullen WR, Dodd M: Appl Organomet Chem 2, 1 (1988)
166. Cappon CJ: LC-GC 5, 400 (1987)
167. Irgolic KJ, Stockton RA, Chakraborti D in: Arsenic, Industrial, Biomedical, and Environmental Perspectives (eds Lederer WH, Fensterheim RJ), Van Nostrand Reinhold, New York 1983, Chap 22, pp 282–308
168. Irgolic KJ, Stockton RA: Marine Chemistry 22, 265 (1987)
169. Braman RS in: Biological and Environmental Effects of Arsenic (ed Fowler BA), Elsevier, Amsterdam 1983, Chap 3
170. Kirchgessner M, Grassmann E, Roth HP, Spoerl R, Schnegg A in: Nuclear Techniques in Animal Production and Health, International Atomic Energy Agency, Vienna 1976, p 61
171. a Anke M, Groppel B, Kronemann H: in Trace element- Analytical Chemistry in Medicine and Biology (ed Braetter P), Vol 3, de Gruyter, Berlin 1984, pp 430–451 and references cited therein
    b Anke M: in Trace Elements in Human and Animal Nutrition (ed W Mertz), Vol 2, Academic Press, New York 1986, pp 347–372 and references cited therein
172. Uthus EO, Cornatzer WE, Nielsen FH: in Arsenic: Industrial, Biomedical and Environmental Perspectives (eds Lederer WH, Fensterheim RJ), Van Nostrand Rheinhold, New York 1983, Part III, pp 173–189
173. Nielsen FH, Uthus EO, Cornatzer WE: Biological Trace Element Research 5, 389 (1983)
174. Nielsen FH, Uthus EO: Biochemistry of the Elements 3 (Biochemistry of the Essential Ultratrace Elements), 319 (1984) and references cited therein
175. Zingaro RA in: Arsenic: Industrial, Biomedical, and Environmental Perspectives (eds. Lederer WH, Fensterheim RJ), Van Nostrand Rheinhold, New York 1983, Part V, pp. 327–347
176. Most K.-H: Ph.D thesis (1939) Graz University, Austria
177. Lenihan J: The Crumbs of Creation, IOP, Bristol 1988, pp 116–123
178. Moncada S, Palmer RM, Higgs EA: Biochem Pharmacol 38, 1709 (1989)
179. Schmidt LH, Anke M, Groppel B: Zentralbl Pharm Pharmakother Laboratoriumdiagn. 127, 197 (1988)
180. Frost DV: Sci Total Environ 28, 455 (1983)
181. Kaemmerer K: Der praktische Tierarzt 4, 380 (1982)
182. Vahter M, Marafante E in: The biological alkylation of heavy elements (eds Craig PJ, Glocking F), Royal Society of Chemistry, London 1988, pp 105–119
183. Thayer JS, Brinckman FE: Adv Organomet Chem 20, 313 (1982)
184. Irgolic KJ in: Frontiers in Bioinorganic Chemistry (ed Xavier AV), VCH, Weinheim 1986, pp 399–408 and references cited therein
185. Aposhian HV in: Reviews in Biochemical Toxicology (eds Hodgson E, Bend J R, Philpot RM), Vol 10, Elsevier, Amsterdam 1989, pp 265–299
186. Challenger F: Chem Rev 36, 315 (1945)
187. Cullen WR, McBride BC, Manji H, Pickett AW, Reglinski J: Appl Organomet Chem 3, 71 (1989)
188. Pickett AW, McBride BC, Cullen WC: Appl Organomet Chem 2, 479 (1988)
189. Irgolic KJ, Banks CH, Bottino NR, Chakraborti D, Gennity JM, Hillman DC, O'Brien DH, Pyles RA, Stockton RA, Wheeler AE, Zingaro RA: NBS Special Publications (U.S.) 618, 244 (1981)
190. Lewin PK, Hancock RGV, Voynovich P: Nature 299, 627 (1982)
191. Buchet JP, Lauwerys R: Toxicology and Applied Pharmacology 91, 65 (1987)

192. Buchet JP, Lauwerys R: Biochem Pharmacol 37, 3149 (1988)
193. Cullen WR, McBride BC, Reglinski J: J Inorg Biochem 21, 179 (1984)
194. Cullen WR, McBride BC, Reglinski J: J Inorg Biochem 21, 45 (1984)
195. Reglinski J, Smith WE, Sturrock RD: Magnetic Resonance in Medicine 6, 217 (1988)
196. Wilson IB in: Molecular Basis of Nerve Activity; Proceedings of the International Symposium in Memory of David Nachmansohn (ed Changeux JP), de Gruyter, Berlin 1985, pp 667–78 and references cited therein
197. Webb JL: Enzymes and Metabolic Inhibitors, Vol 3, Academic Press, London 1966, Chap 6, pp 595–793 and references cited therein
198. Oehme FW in: Toxicity of Heavy Metals in the Environment (ed Oehme FW), Marcel Dekker, New York 1979, Part 2, Chap 31, pp 945–952
199. Dill K, Adams ER, O'Connor RJ, Chong S, McGown EL: Arch Biochem Biophys 257, 293 (1987)
200. Kobayashi JH, Yuyama A, Ishihara M, Matsusaka N: Neuropharmacology 26, 1707 (1987)
201. Bernier M, Laird DM, Lane MD: J Biol Chem 263, 13626 (1988)
202. Steinhelper ME, Olson MS: Biochem Pharmacology 37, 1167 (1988)
203. Pressman D, Sternberger LA: J Immunology 66, 609 (1951)
204. Grossberg AL, Pressman D: Biochemistry 7, 272 (1968)
205. Freedman MH, Grossberg AL, Pressman D: J Biol Chem 243, 6186 (1968)
206. Freedman MH, Grossberg AL, Pressman D: Biochemistry 7, 1941 (1968)
207. Milner ECB, Capra JD: J Immunology 129, 193 (1982)
208. Rathbun G, Sanz I, Meek K, Tucker P, Capra JD: Adv Immunology 42, 95 (1988)
209. Goodman JW: Annales de L'Institut Pasteur 137D, 303 (1986) and references cited therein
210. Bullock WW: Immunological Rev 39, 3 (1978)
211. Alkan SS in: The Immune System (eds Steinberg CM, Lefkovits I), Vol. II, Karger, Basel 1981, pp 329–335
212. Goodman JW, Fong S, Lewis GK, Kamin R, Nitecki DE, Der Balian G: Immunological Rev 39, 36 (1978)
213. (a) Barber HJ: J Chem Soc., 1020 (1929); (b) Cohen A, King H, Strangeways WI: J Chem Soc, 2505 (1932)
214. Blakley BR, Sisoda CS, Mukkur TK: Toxicol Appl Pharmacol 52, 245 (1980)
215. Bauer H: Annals New York Acad Sci 59, 150 (1954)
216. Schwyzer R: Angew Chem 66, 345 (1954)
217. Holmstedt B, Liljestrand G: Readings in Pharmacology, Pergamon, Oxford 1963, pp 281–293
218. Cullen WR in: Dictionary of Organometallic Compounds (ed Buckingham J), Chapman and Hall, London 1984, pp 123–187, (As-00152, As-00093, As-00106, 00107)
219. Abernathy JR in: Arsenic, Industrial, Biomedical, and Environmental Perspectives (eds. Lederer WH, Fensterheim RJ), Van Nostrand Reinhold, New York 1983, Chap 5, pp 57–62
220. Alden JC in: Arsenic, Industrial, Biomedical, and Environmental Perspectives (eds. Lederer WH, Fensterheim RJ), Van Nostrand Reinhold, New York 1983, Chap 6, pp 63–71
221. Woolson EA in: Biological and Environmental Effects of Arsenic (ed Fowler BA), Elsevier, Amsterdam 1983, Chap 2, pp 51–139
222. Dictionary of Organometallic Compounds (ed Macintyre JE), 1st Supplement, Chapman and Hall, London 1985, pp 19–24, (As-10002)
223. Nagasawa M: Chem Abs 56, 10637 (1962)
224. Yeager CC: Chem Abs 66, 86431 (1967)
225. Wang CS, McGee TW: Chem Abs 74, 64308 (1971); Wang C.-S, McGee TW: Chem Abs 76, 127155 (1972)
226. Knowles FC, Benson AA: Plant Physiol 71, 235 (1983)
227. Anderson CE in: Arsenic, Industrial, Biomedical, and Environmental Persp ectives (eds Lederer WH, Fensterheim RJ), Van Nostrand Reinhold, New York 1983, Chap 8, pp 89–97 and references cited therein
228. Ledet AE, Buck WB in: Toxicity of Heavy Metals in the Environment (ed Oehme FW), Marcel Dekker, New York 1979, Part One, Chap 17, pp 375–391
229. Hedlund B, Norin H, Christakopoulos A, Alberts P, Bartfai T: Journal of Neurochemistry 39, 871 (1982)
230. Silver S in: Environmental Speciation and Monitoring Needs for Trace Metal-Containing Substances from Energy-Related Processes (eds Brinckman FE, Fish RH), 1981, National Bureau of Standards Special Publ 618, Washington DC, pp 301–319

231. Silver S in: Staphylococci and Staphylococcal Infections, Zbl Bakt Suppl 10 (ed Jeljaszewicz J), 1981, Gustav Fischer, Stuttgart, pp 589–599
232. Thiel T: J Bacteriol *170*, 1143 (1988)
233. Silver S (1984) in: Nriagu JO (ed) Changing metal cycles and human health, Springer, Berlin Heidelberg New York, p 199
234. Silver S, Nucifora G, Chu L, Misra TK: Trends in Biochemical Science *14*, 76 (1989) and references cited therein
235. Rosen BP, Ambudkar SV, Borbolla MG, Chen CM, Mobley HLT, Tsujibo H, Zlotnick GW in: Prog Clin Biol Res, vol 168 (Epithelial Calcium and Phosphate Transport: Molecular and Cellular Aspects, eds Bronner F, Peterlik M), AR Liss, New York 1984, pp 219–226
236. Rosen BP, Weigel U, Karkaria C, Gangola P in: The Ion Pumps: Structure, Function, and Regulation (ed Stein WD), AR Liss, New York 1988, pp 105–112
237. Rosen BP, Weigel U, Karkaria C, Gangola P: J Biol Chem *263*, 3067 (1988)
238. Tisa LS, Rosen BP: J Biol Chem *265*, 190 (1990)
239. Marcovich D, Tapscott RE: J Amer Chem Soc *102*, 5712 (1980) and references cited therein
240. Banks CH, Daniel JR, Zingaro RA: J Med Chem *22*, 572 (1979)
241. Kenney LJ, Kaplan JH: J Biol Chem *263*, 7954 (1988)
242. Forbush B (III): J Biol Chem *263*, 7961 (1988)
243. Pipkin JL, Anson JF, Hinson WG, Burns ER, Casciano DA, Sheehan DM: Biochim Biophys Acta *927*, 334 (1987)
244. Pershagen G in: Environmental Carcinogens, Selected Methods of Analysis, Vol 8, International Agency for Research on Cancer Scientific Publications, Lyons 1986, Chap 3, pp. 45–61
245. Pershagen G: Environmental Health Perspectives *40*, 93 (1981)
246. Axelson A: J Toxicol Environ Health *6*, 1229 (1980)
247. Léonard A, Lauwerys RR: Mutation Research *75*, 49 (1980)
248. Wright D: Medical History *27*, 184 (1983)
249. Jacobson-Kram D, Montalbano D: Environmental Mutagenesis *7*, 787 (1985)
250. Zingaro RA: Chemica Scripta *8A*, 51 (1975)
251. Haiduc I, Silvestru C, Coord Chem Rev *99*, 253 (1990) and references cited therein.
252. Mirabelli CK, Hill DT, Faucette LF, McCabe FL, Girard GR, Bryan DB, Sutton BM, O'Leary Bartus J, Crooke ST, Johnson RK: J Med Chem *30*, 2181 (1987)
253. Mirabelli CK, Johnson RK, Hill DT, Faucette LF, Girard GR, Kuo GY, Sung CM, Crooke ST: J Med Chem *29*, 218 (1986)
254. Jarrett PS, Ni Dhubhghaill OM, Sadler PJ, unpublished results
255. Berners-Price SJ, Sadler PJ: Struct Bonding *70*, 27 (1988) and references cited therein
256. Hoke GD, Rush GF, Bossard GE, McArdle JV, Jenson BD, Mirabelli CK: J Biol Chem. *263*, 11203 (1988)
257. Kirschner S, Wei YK, Francis D, Bergman JG: J Med Chem *9*, 369 (1966)
258. Köpf-Maier P, Klapötke T: Inorg Chim Acta *152*, 49 (1988)
259. Gupta P, O'Marro S, Rinaldo CR, Ho M: 5th International Conference on AIDS, Montreal 1989, Section C, M.C.P 133.
260. Jasmin C, Chermann JC, Herve G, Teze A, Souchay P, Boy-Loustau C, Raybaud N, Sinoussi F, Raynaud M: J Nat Cancer Inst *53*, 469 (1974)
261. Edmonds JS, Francesconi KA, Cannon JR, Raston CL, Skelton BW, White AH: Tet Lett *18*, 1543 (1977)
262. Edmonds JS, Francesconi KA: Nature *289*, 602 (1981)
263. Nisizawa K, Noda H, Kikuchi R, Watanabe T: Hydrobiologia *151–152*, 5 (1987)
264. Maher W, Butler E: Applied Organometallic Chemistry *2*, 191 (1988) and references cited therein
265. Edmonds JS, Francesconi KA: Appl Organometallic Chemistry *2*, 297 (1988)
266. Edmonds JS, Francesconi KA: Experientia *43*, 553 (1987)
267. Phillips DJH, Depledge MH: Water Science Technology *18*, 213 (1986)
268. Edmonds JS, Francesconi KA: J Chem Soc Perkin Trans I, 2375 (1983)
269. Edmonds JS, Francesconi KA in: The biological alkylation of heavy elements (eds Craig PJ, Glocking F), Royal Society of Chemistry, London 1988, pp 138–141 and references cited therein
270. Edmonds JS, Francesconi KA, Hansen JA: Experientia *38*, 643 (1982)
271. Norin H, Christakopoulos A: Chemosphere *11*, 287 (1982)
272. Hanaoka K, Yamamoto H, Kawashima K, Tagawa S, Kaise T: Appl Organomet Chem. *2*, 371 (1988)

273. Maher WA: Biological Trace Element Research *6*, 159 (1984)
274. Vahter M, Marfante E, Dencker L: Science of the Total Environment *30*, 197 (1983)
275. Cannon JR, Saunders JB, Toia RF: Science of the Total Environment *31*, 181 (1983)
276. Irvin TR, Irgolic KJ: Applied Organometallic Chemistry *2*, 509 (1988)
277. Yancy PH, Clarke ME, Hand SC, Bowlus RD, Somero GN: Science *217*, 1214 (1982)
278. Dévai I, Felföldy L, Wittner I, Plósz S: Nature *333*, 343 (1988)
279. Berners-Price SJ, Mirabelli CK, Johnson RK, Mattern MR, McCabe FL, Faucette LF, Sung C-H, Mong S-M, Sadler PJ, Crooke ST: Cancer Research *46*, 5486 (1986)